T0242860

Lecture Notes in Mathematics

Edited by J.-M. Morel, F. Takens and B. Teissier

Editorial Policy for Multi-Author Publications: Summer Schools / Intensive Courses

1. Lecture Notes aim to report new developments in all areas of mathematics and their applications – quickly, informally and at a high level. Mathematical texts analysing new developments in modelling and numerical simulation are welcome. Manuscripts should be reasonably self-contained and rounded off. Thus they may, and often will, present not only results of the author but also related work by other people. They should provide sufficient motivation, examples and applications. There should also be an introduction making the text comprehensible to a wider audience. This clearly distinguishes Lecture Notes from journal articles or technical reports which normally are very concise. Articles intended for a journal but too long to be accepted by most journals, usually do not have this „lecture notes" character.

2. In general SUMMER SCHOOLS and other similar INTENSIVE COURSES are held to present mathematical topics that are close to the frontiers of recent research to an audience at the beginning or intermediate graduate level, who may want to continue with this area of work, for a thesis or later. This makes demands on the didactic aspects of the presentation. Because the subjects of such schools are advanced, there often exists no textbook, and so ideally, the publication resulting from such a school could be a first approximation to such a textbook.

   Usually several authors are involved in the writing, so it is not always simple to obtain a unified approach to the presentation.

   For prospective publication in LNM, the resulting manuscript should not be just a collection of course notes, each of which has been developed by an individual author with little or no co-ordination with the others, and with little or no common concept. The subject matter should dictate the structure of the book, and the authorship of each part or chapter should take secondary importance. Of course the choice of authors is crucial to the quality of the material at the school and in the book, and the intention here is not to belittle their impact, but simply to say that the book should be planned to be written by these authors jointly, and not just assembled as a result of what these authors happen to submit.

   This represents considerable preparatory work (as it is imperative to ensure that the authors know these criteria before they invest work on a manuscript), and also considerable editing work afterwards, to get the book into final shape. Still it is the form that holds the most promise of a successful book that will be used by its intended audience, rather than yet another volume of proceedings for the library shelf.

3. Manuscripts should be submitted (preferably in duplicate) either to Springer's mathematics editorial in Heidelberg, or to one of the series editors (with a copy to Springer). Volume editors are expected to arrange for the refereeing, to the usual scientific standards, of the individual contributions. If the resulting reports can be forwarded to us (series editors or Springer) this is very helpful. If no reports are forwarded or if other questions remain unclear in respect of homogeneity etc, the series editors may wish to consult external referees for an overall evaluation of the volume. A final decision to publish can be made only on the basis of the complete manuscript; however a preliminary decision can be based on a pre-final or incomplete manuscript. The strict minimum amount of material that will be considered should include a detailed outline describing the planned contents of each chapter.

   Volume editors and authors should be aware that incomplete or insufficiently close to final manuscripts almost always result in longer evaluation times. They should also be aware that parallel submission of their manuscript to another publisher while under consideration for LNM will in general lead to immediate rejection.

Continued on inside back-cover

Lecture Notes in Mathematics

1900

Editors:
J.-M. Morel, Cachan
F. Takens, Groningen
B. Teissier, Paris

Erwin Bolthausen, Anton Bovier (Eds.)

Spin Glasses

 Springer

Editors

Erwin Bolthausen
Mathematics Institute
University of Zürich
Winterthurerstraße 190
8057 Zürich, Switzerland
e-mail: eb@math.unnizh.ch

Anton Bovier
Weierstraß Institute for Applied Analysis and Stochastics
Mohrenstraße 39
10117 Berlin
Germany

and

Institute for Mathematics
Berlin University of Technology
Straße des 17. Juni 136
10623 Berlin
Germany
e-mail: bovier@wias-berlin.de

Library of Congress Control Number: 2006932484

Mathematics Subject Classification (2000): 82B44, 82C44, 82B20, 60K35, 60K37

ISSN print edition: 0075-8434
ISSN electronic edition: 1617-9692

ISBN 978-3-540-40902-1 Springer Berlin Heidelberg New York

DOI 10.1007/3-540-40902-5

Springer is a part of Springer Science+Business Media
springer.com
© Springer-Verlag Berlin Heidelberg 2007

Typesetting by the editors and SPi using a Springer LATEX package
Cover design: WMXDesign GmbH, Heidelberg

Printed on acid-free paper   SPIN: 11859840   VA41/3100/SPi   5 4 3 2 1 0

Preface

Spin glasses have become a paradigm for highly complex disordered systems. In the 1960'ies, certain magnetic alloys were found to have rather anomalous magnetic and thermal properties that seemed to indicate the existence of a new kind of phase transition, clearly distinct from conventional ferromagnetic materials. The origin of these anomalies was soon deemed to lie in two features: the presence of competing signs in the two-body interactions, and the disorder in the positions of the magnetic atoms in the alloy. This has led to the modelling of such materials in the form of spin-systems with random interactions. In the 1970ies, two principle models were proposed: the Edwards-Anderson model, which is a lattice spin system with random nearest neighbor interactions and as such is the randomized version of the classical Ising model; and the Sherrington-Kirkpatrick model, proposed as a mean field model, where all spins interact with each other on equal footing, which is a randomized version of the Curie-Weiss model. The SK-model was clearly intended to provide a simple, solvable caricature of the Edwards-Anderson model, that should give some insights into the nature of the spin glass transitions, just as the Curie-Weiss model allows a partial understanding of ferromagnetic phase transitions. The remarkable interest that the spin glass problem has received is largely due to the fact that neither of the two models turned out to be easily tractable. The Sherrington-Kirkpatrick model was solved on a heuristic level through the remarkable "replica symmetry breaking" ansatz of Parisi, which not only involved rather unconventional mathematical concepts, but also exhibited that the thermodynamic limit of this model should be described by an extraordinarily complex structure. The short-range Edwards Anderson model has been even more elusive, and beyond some rather rudimentary rigorous results, most of our insight into the model is based on numerical simulations, which in themselves prove to be a highly challenging task.

Mathematicians became interested in this problem in the late 1980ies, but on a larger scale in the 1990ies, starting with work of Pastur and Shcherbina, and the systematic programmes initiated by Guerra on the one hand and Talagrand on the other. In 1996 a workshop in Berlin brought together the leading experts in the field. The state of the art at that time is to a large extend documented in the volume "Mathematical Aspects of Spin Glasses and Neural Networks", edited by A. Bovier and P. Picco (Birkhäuser, 1997). Since then, the progress made in the field has exceeded all expectations. Even as we began planning for a new workshop on the mathematics of spin glasses that was finally held at the Centro Stefano Franscini on the Monte Verità, we did not anticipate that he timing of the event would allow to present for the first time some ground breaking progress. In 2002, Francesco Guerra published an upper bound on the free energy of the SK model that coincided with the Parisi solution. This was the first time that this remarkable construction was to be related to a mathematically rigorous result. Less than a year later, Michel Talagrand announced that he could prove the corresponding lower bound, thus establishing the Parisi solution in a fully rigorous manner.

The Monte Verità meeting thus fell into a most exciting period. It was attended by most of the leading experts on spin glasses, including David Sherrington, Giorgio Parisi, Francesco Guerra, Michel Talagrand, Michael Aizenman, Chuck Newman, and Daniel Stein, to name a few. Besides the reports on the progress mentioned above, the participants and invited speakers reported on a wealth of interesting new results around spin glasses, both on the static and dynamic aspect. As a result we decided to collect a number of invited review papers to document the state of the art in spin glass theory today. The result of this is the present book. It contains a general introduction to the spin glass problem, written by E. Bolthausen, that will serve in particular as a pedagogical guide to the description of the nature of the Parisi solution and the derivation of Guerra's bound in the formulation of Aizenman, Sims, and Star. A. Bovier and I. Kurkova shed light on the Parisi solution from another angle by deriving and describing the asymptotics of the Gibbs measure in another class of spin glass models, the Generalized Random Energy models, in full detail. D. Sherrington gives an account of the history of the spin glass problem from a more physical perspective. M. Talagrand's contribution is a pedagogical presentation of his celebrated proof of the validity of the Parisi solution. Two articles by Ch. Newman and D. Stein discuss the latest developments in the ongoing dispute on the question, whether the predictions of the mean field Sherrington-Kirkpatrick model have any implications for the behavior of short range spin glasses. Finally, A. Guionnet gives an account of what has been achieved in the understanding of another outstanding issue about spin glasses, namely their non-equilibrium properties.

We hope that this volume will serve as a reference handbook for anyone wanting to get an idea of where we are in the theory of spin glasses, and what this subject is all about.

Erwin Bolthausen
Anton Bovier

Contents

Mean Field Models for Spin Glasses: Some Obnoxious Problems

Much Ado about Derrida's GREM

Random Media and Spin Glasses: An Introduction into Some Mathematical Results and Problems

Erwin Bolthausen

Mathematics Institute, University of Zürich
Winterthurerstraße 190, 8057 Zürich, Switzerland
e-mail: eb@math.unnizh.ch

1 Introduction

No materials in the history of solid state physics have been as intriguing and perplexing than certain alloys of.ferromagnets and conductors, such as AuFe or CuMg, known as *spin glasses*. The attempts to model these systems have led to a class of *disordered spin systems* whose mathematical analysis has proven to be among the most fascinating fields of statistical mechanics over the last 25 years. Even the seemingly most simple model class, the *mean-field models* introduced by Sherrington and Kirkpatrick [1] now known as SK-models have proven to represent an amazingly rich structure that is mathematically extraordinarily hard to grasp. Theoretical physics has produced an astounding solution describing the thermodynamics properties of these models that is based on ad hoc constructions (so-called "replica symmetry breaking" [2]) that so far have largely resisted attempts to be given a concrete mathematical sense. From a purely mathematical point of view, the problem posed here represents a canonical problem in the theory of stochastic processes in high dimensions and as such the interest in it transcends largely the original physical question. The fact that the heuristic approach of theoretical physics, if given a clear mathematical meaning, would give a totally new and powerful tool for the analysis of such questions is the reason why there has been a strong upsurge in interest from within the mathematical, and in particular probabilistic community in this and related problems. Moreover, the same types of mathematical problems arise in many areas of applications that are of great current interest. For example, heuristic methods of statistical mechanics make powerful prediction concerning numerous problems of combinatorial optimization, computer science, and information technology.

For a long time progress on mathematically rigorous results in this field have been extremely limited, but over the last years the situation has changed considerably due to the results of Bovier, Comets, Derrida, Gayrard,

Newman, Pastur, Picco, Shcherbina, Stein, Talagrand, and Toninelli, for instance. Michel Talagrand has developed in a systematic way an induction technique known as the "cavity method" as a tool to analyze in a rigorous way random Gibbs measures. This has allowed him to confirm predictions made by the heuristic "replica method" mostly in domains where the so-called replica-symmetric solution is predicted to hold; in mathematical terms, this corresponds to situations where the Gibbs measure is asymptotically a (random) product measure. The cavity method then allows to precisely compute the corresponding parameters. Interestingly, the method can also be applied in some situations where the Gibbs measure is a nontrivial mixture of product measures ("one-step replica symmetry breaking"), such as the p-spin Sherrington–Kirkpatrick model. Much of this can be found in Talagrand's book [3].

Another discovery was made by Guerra and Ghirlanda. This concerns a set of recursive relations between so-called multioverlap distributions. In certain cases it could be shown that they determine a universal structure in the Gibbs measures of these systems. In particular, these identities proved crucial in the work of Talagrand on the p-spin models, and in recent work of Bovier and Kurkova who used them to prove convergence and describe the limit of the Gibbs measures in a class of models introduced by Derrida, the so-called generalized random energy models.

The most spectacular successes recently, initiated by Francesco Guerra, are coming from interpolation techniques between different processes. Such methods are in principle well established in the analysis of Gaussian processes. Nonetheless, their judicious use has led to very remarkable results: Guerra and Toninelli [4] used them to prove the existence of the limit of the free energy in the SK (and many similar) models. A bit later, Guerra [5] has been able to prove that the predicted expression for the free energy of the SK-model from replica theory is at least a lower bound, and finally, Talagrand [6] has been able to refine the technique and combine it with his cavity method to control the error in Guerra's bound, and in this way he proved the Parisi formula [7] in the full temperature regime of the SK-model. Despite of these successes, there still remain many open problems, and it is perhaps fair to say that even the SK-model, where the Parisi formula has now been proved, is still very poorly understood. For instance, an understanding of the so-called ultrametricity is completely lacking, although it is at the very heart of the physics theory of the model. Even more importantly, there are many models where interpolation techniques had been far less successful, and where our understanding is till restricted to the part of the parameter space outside the spin glass phase.

In a second development, the analysis of the stochastic dynamics of highly disordered model is starting to make progress. Important contributions are due to Ben Arous and Guionnet and Grunwald, who derived asymptotic dynamics in Langevin and Glauber dynamics of the SK-model. Spin glass dynamics is supposed to have so-called "aging" which means that the systems react slower the older it gets. There are mathematically precise descriptions of this behavior which, however, have not yet been proved for the SK-model, but for simpler models there has been a lot of progress, recently (see the paper of Alice Guionnet in this volume).

In this introductory notes, I will give an overview for some of the developments, but I will mainly focus on mathematical results, and on results which are developed later in other contributions in this volume in more details.

For an overview over recent developments and perspectives in physics, see the article of Sherrington [8] in this volume. A topic which I leave out in this introduction is short-range spin glass models. This is presently still quite a controversial subject, even in the physics literature. In recent years, Newman and Stein [9, 10] have obtained results.

The focus given here in my introductory notes is on the mean-field model, and in particular on the recent mathematical developments around Guerra's interpolation technique (Sects. 4.1 and 5.3), the Talagrand's version of the cavity method in Sect. 4.2, and the random energy models in Sect. 5. For more on this subject, see the article by Bovier and Kurkova [11]. For a more in depth overview of Talagrand's application of the Guerra interpolation to SK, and its combination with the cavity method, see his article in this volume [12]. A topic I only shortly mention here is the dynamical behavior of spin glasses which is presented in much more depth by Guionnet [13].

2 The Basic Mathematical Models

The usual lattice spin–models of (nonrandom) Ising type are defined as follows. Consider a finite set Λ, and let $\Sigma_\Lambda \overset{\text{def}}{=} \{-1, 1\}^\Lambda$. Let further $A = (a_{ij})_{i,j \in \Lambda}$ be a real symmetric matrix, and $\mathbf{h} = (h_i)_{i \in \Lambda}$ be a real vector. The Hamiltonian with these parameters is the mapping $H_{A,\mathbf{h}} : \Sigma_\Lambda \to \mathbb{R}$ defined by

$$-H_{A,\mathbf{h}}(\sigma) \overset{\text{def}}{=} \frac{1}{2} \sum_{i,j \in \Lambda} a_{ij} \sigma_i \sigma_j + \sum_{i \in \Lambda} h_i \sigma_i,$$

and the Gibbs measure $\mathcal{G}_{A,\mathbf{h}}$ on Σ_Λ is defined by

$$\mathcal{G}_{\Lambda,A,\mathbf{h}}(\sigma) \overset{\text{def}}{=} \frac{1}{Z_{\Lambda,A,\mathbf{h}}} \exp\left[-H_{A,\mathbf{h}}(\sigma)\right], \qquad (2.1)$$

where of course

$$Z_{\Lambda,A,\mathbf{h}} \overset{\text{def}}{=} \sum_\sigma \exp\left[-H_{A,\mathbf{h}}(\sigma)\right] \qquad (2.2)$$

is the so-called partition function, the normalizing factor in order that the Gibbs distribution is a probability distribution. (In the physics literature, one takes the Hamiltonian with a minus sign, so I keep with this tradition, although it is mathematically a bit annoying.) Of great importance is the (finite volume) free energy, defined by

$$F_\Lambda (A, \mathbf{h}) = \frac{1}{|\Lambda|} \log Z_{\Lambda, A, \mathbf{h}}. \qquad (2.3)$$

The importance of this quantity is coming from the fact that most of the physical interesting quantities can be expressed through it, like mean magnetization, entropy, etc. For instance

$$\frac{\partial}{\partial h_j} F_\Lambda (A, \mathbf{h}) = \frac{1}{|\Lambda|} \sum_\sigma \sigma_j \mathcal{G}_{\Lambda, A, \mathbf{h}} (\sigma),$$

and summing over $j \in \Lambda$ gives the mean magnetization under the Gibbs measure.

As for a finite set Λ, detailed properties of such models are usually impossible to describe, one usually tries to perform the "thermodynamic limit." For instance, if $\Lambda \subset \mathbb{Z}^d$, one can usually prove that the limiting free energy

$$f (A, \mathbf{h}) \stackrel{\text{def}}{=} \lim_{\Lambda \uparrow \mathbb{Z}^d} F_\Lambda (A, \mathbf{h})$$

exists, provided the Λ approach $\mathbb{Z}^d$ in a not too nasty way, and A and $\mathbf{h}$ are defined on the whole of $\mathbb{Z}^d$.

The best known example is the Ising model where Λ is a finite (large) box in $\mathbb{Z}^d$, and

$$a_{ij} \stackrel{\text{def}}{=} \begin{cases} \beta \, |i - j| = 1 \\ 0 \text{ otherwise} \end{cases},$$

β the so-called inverse temperature.

Short-range models are usually rather difficult to analyze, and often a qualitatively good approximation is obtained from *mean-field models* where every spin interacts with any other one on equal footing. The simplest mean-field model is the *Curie–Weiss model*. Here

$$a_{ij} \stackrel{\text{def}}{=} \beta / |\Lambda|, \ \forall i, j \in \Lambda.$$

In that case one has with $N \stackrel{\text{def}}{=} |\Lambda|$

$$\frac{1}{2} \sum_{i, j \in \Lambda} a_{ij} \sigma_i \sigma_j = \frac{\beta}{2N} \left\{ \sum_{i \in \Lambda} \sigma_i \right\}^2,$$

and anything one wants to know can be derived from the Stirling approximation, and it becomes an easy exercise in elementary probability. If k has the same parity as N, then

$$\#\left\{\sigma:\sum_{i\in\Lambda}\sigma_i=k\right\}=\frac{N!}{((N+k)/2)!\,((N-k)/2)!}$$

$$\simeq\left(\frac{N}{(N+k)/2}\right)^{(N+k)/2}\left(\frac{N}{(N-k)/2}\right)^{(N-k)/2}$$

$$=2^N\exp\left[-N\left(\frac{1+\frac{k}{N}}{2}\log\left(1+k/N\right)\right.\right.$$

$$\left.\left.+\frac{1-\frac{k}{N}}{2}\log\left(1-\frac{k}{N}\right)\right)\right].$$

From that one sees by a simple Laplace approximation that for constant $h_i=h$, one has that the limiting free energy of the Curie–Weiss model is given by

$$f\left(\beta,h\right)=\log 2+\sup_{t\in[-1,1]}\left[\frac{\beta t^2}{2}+h-\frac{1+t}{2}\log\left(1+t\right)+\frac{1-t}{2}\log\left(1-t\right)\right].$$

In order to appreciate the simplification obtained by the mean-field ansatz, one has to compare that with the tremendously more difficult analysis in the ordinary Ising model as they can be found in standard textbooks, see, e.g., [14]. One aspect, one has however to keep in mind, is that for mean-field models it is difficult to talk about limiting Gibbs measures, "pure states," and the like. This aspect seems to have played a considerable role in the discussions and controversies whether mean-field spin glasses share some properties with short-range spin glasses. As I am not very knowledgeable on this subject, I do not want to comment about this issue, and rather advise the reader to read the contributions of Newman and Stein in this volume.

Spin glasses are models where the interactions are "disordered," which typically means that they are obtained as a random object. A topic which is still very poorly understood is the case of *short-range* random interactions, for instance when $\Lambda=\{-n,\ldots,n\}^d$, and the a_{ij} are independent Gaussians for $|i-j|=1$, and 0 otherwise. This is the *Edwards–Anderson model* on which there are ongoing controversial discussions in the physics community, the more so as it is very difficult to simulate it on computers with a reasonably large box and in interesting dimensions. One of the key issues is the presence of so-called "frustrations." This means that for three sites i,j,k, the interactions between i and j and between j and k may be positive, but between i and k negative. In particular, in contrast to the Ising model, spin glasses usually do not satisfy any of the well-known correlation inequalities, like the FKG inequality.

The situation is considerably better understood for the random field Ising model, where the interactions a_{ij} are the same as for the Ising model, but where the h_i are independent Gaussian random variables. On this, there are now classical results [15, 16], but we will not enter into this subject in this volume.

3 The Sherrington–Kirkpatrick Model

The Sherrington–Kirkpatrick model has the "mean-field" random Hamiltonian

$$-H_{N,\beta,h,\omega} \stackrel{\text{def}}{=} \beta X_{n,\omega}(\sigma) + h \sum_{i=1}^{N} \sigma_i,$$

where

$$X_{n,\omega}(\sigma) \stackrel{\text{def}}{=} \frac{1}{\sqrt{N}} \sum_{1 \le i < j \le N} J_{ij}(\omega) \sigma_i \sigma_j,$$

and the J_{ij} are independent standard Gaussian random variables, defined on some probability space $(\Omega, \mathcal{F}, \mathbb{P})$. (I will constantly use $\mathbb{P}$ for the probability measure governing the disorder, with $\mathbb{E}$ as the corresponding expectation.) We will often drop ω and N in such expressions. One first observes that the $1/\sqrt{N}$-normalization is the right one in order to catch the "spirit" of a mean-field interaction: The total influence of the spins σ_j, $j \ne i$, on the ith spin is

$$\frac{1}{\sqrt{N}} \sum_{j > i} J_{ij}\sigma_j + \frac{1}{\sqrt{N}} \sum_{j < i} J_{ji}\sigma_j$$

which is of order 1.

Remark that for any σ, $X_N(\sigma)$ is a random variable, and indeed a centered Gaussian one. The covariances are given by

$$\mathbb{E}(H_N(\sigma) H_N(\sigma')) = \frac{1}{N} \sum_{1 \le i < j \le N} \sigma_i \sigma_j \sigma_i' \sigma_j' = \frac{1}{2N} \sum_{i,j=1}^{N} \sigma_i \sigma_j \sigma_i' \sigma_j' - \frac{1}{2}$$

$$= \frac{N}{2} \left(\frac{1}{N} \sum_{i=1}^{N} \sigma_i \sigma_i' \right)^2 - \frac{1}{2}. \tag{3.1}$$

The quantity in brackets is the so-called *overlap* of the two spin configurations

$$R_N(\sigma, \sigma') \stackrel{\text{def}}{=} \frac{1}{N} \sum_{i=1}^{N} \sigma_i \sigma_i'.$$

The (random) Gibbs distribution $\mathcal{G}_{N,\beta,h,\omega}$, the partition function $Z_{N,\beta,h,\omega}$, and the (finite N) free energy $F_{N,\beta,h,\omega}$ are defined as in (2.1)–(2.3).

There exist also variants where h is a random variable or where $h \sum_{i=1}^{N} \sigma_i$ is replaced by $\sum_{i=1}^{N} h_i \sigma_i$, where the h_i are random variables, e.g., $h_i = \gamma g_i + h$, $\gamma > 0$, $h \in \mathbb{R}$, and the g_i again being independent standard Gaussian random variables. This generalization is actually important, because the more general version appears naturally in the interpolation scheme invented by Guerra (see Sect. 4.1).

The free energy F_N is still a random variable, and we write

$$f_N(\beta, h) \stackrel{\text{def}}{=} \mathbb{E} F_{N,\beta,h},$$

the so-called "quenched" free energy. Sometimes, "quenched" refers to the random quantity only, but there is not much difference, as we will explain. In contrast, the so-called "annealed" free energy is obtained by taking the expectation inside the logarithm. By Jensens's inequality, f_N is dominated by the annealed free energy.

The model is evidently closely connected with questions probabilists have been interested in for a long time, namely maxima (or minima) of (Gaussian) random vectors. For instance, $\lim_{\beta \to \infty} (1/\beta) \log Z_{N,\beta,0}$ is simply $\max_\sigma H_N(\sigma)$, which is just the maximum of a family of correlated Gaussians with a simple covariance structure. Probabilists have developed methods to investigate such questions for a long time, e.g., Dudley, Fernique, Talagrand, and many others. It is not difficult to see that $\max_\sigma H_N(\sigma)$ is of order N and to prove that there are constants $0 < C_1 < C_2$ satisfying

$$\lim_{N \to \infty} \mathbb{P}\left(C_1 N \leq \max_\sigma H_N(\sigma) \leq C_2 N\right) = 1.$$

However, the standard probabilistic techniques cannot derive the exact constant, which the Parisi theory does, revealing a marvelous mathematical structure behind which is still *very* poorly understood, to this day.

3.1 Basic Properties of the SK-Model

The first question one typically answers is the existence of the free energy in the thermodynamical limit (here just $N \to \infty$). It is, however, not at all clear that the free energy

$$\lim_{N \to \infty} F_N(\beta, h)$$

exists. In principle, even if the limit exists, it .could be a random variable. This possibility is, however, ruled out by Gaussian concentration inequalities. One says that the free energy is "self-averaging," meaning that no randomness remains in the $N \to \infty$ limit. For a proof of the following inequality, see for instance [17].

Proposition 3.1. *Let γ_n be the standard Gaussian distribution on $\mathbb{R}^n$. Let $f : \mathbb{R}^n \to \mathbb{R}$ be a Lipshitz continuous function with Lipshitz constant 1. Then for any $u > 0$*

$$\gamma_n\left(f > \int f d\gamma_n + u\right) \leq \exp\left[-u^2/2\right].$$

If we apply this inequality to $F_N(\beta, h)$, regarded as a function of the standard Gaussian vector $(J_{ij})_{1 \leq i < j \leq N}$, then one gets

$$\mathbb{P}\left(\left|\frac{1}{N} \log Z_{N,\beta,h} - \frac{1}{N}\mathbb{E}\log Z_{N,\beta,h}\right| \geq N^{-1/4}\right) \leq 2\exp\left[-\frac{N^{1/2}}{\beta^2}\right].$$

It is therefore clear that instead of investigating $\lim_{N \to \infty} F_N(\beta, h)$, one can as well investigate the nonrandom object $\lim_{N \to \infty} f_N(\beta, h)$. The existence of this limit had been open for a long time, until Guerra and Toninelli [4] found a very nice, and not so obvious superadditivity property

$$\mathbb{E}\log Z_{N_1+N_2} \geq \mathbb{E}\log Z_{N_1} + \mathbb{E}\log Z_{N_2}, \tag{3.2}$$

from which one easily derives that

$$f(\beta, h) = \lim_{N \to \infty} f_N(\beta, h)$$

exists.

For the SK-model, the inequality came somewhat as a surprise. The proof is by a simple but very clever interpolation scheme which interpolates between the $(N_1 + N_2)$ system and the two independent smaller systems. Such interpolation schemes are at the very base of the recent progress in the understanding of the SK-model, as we will see later.

There are many quantities in the SK-model which are *not* self-averaging in the $N \to \infty$ limit, i.e., which stay random (or at least are believed to be so). An example is the overlap of two independent "replicas." Take σ, σ' to be two independent realizations under $\mathcal{G}_{N,\beta,h,\omega}$ for a fixed ω, and calculate $R_N(\sigma, \sigma')$, and then take the Gibbs expectation. This is still a random variable (being a function of the interaction strengths). For small β, these random variables have a nonrandom limit for $N \to \infty$, but the limit stays random for large β. The case $h = 0$ has some evident symmetry properties which make life easier, particularly in the high-temperature region.

For $h = 0$ and small enough β, the ("quenched") free energy equals the "annealed" free energy, a fact first proved in [18, 19].

Theorem 3.2. *For $h = 0$, and $\beta \leq 1$, one has*

$$f(\beta) = \lim_{N \to \infty} \frac{1}{N} \log \mathbb{E} Z_{N,\beta} = \frac{\beta^2}{4} + \log 2. \tag{3.3}$$

The second equation is evident

$$\mathbb{E} Z_{N,\beta} = \sum_{\sigma} \mathbb{E}\exp\left[\beta H_N(\sigma)\right] = \sum_{\sigma} \exp\left[\frac{\beta^2}{2}\operatorname{var}(H_N(\sigma))\right]$$
$$= 2^N \exp\left[\frac{\beta^2}{2}\operatorname{var}(H_N(\sigma))\right] = 2^N \exp\left[\frac{\beta^2}{2}\left(\frac{N}{2} - \frac{1}{2}\right)\right]$$

from which the claim follows. The somewhat astonishing fact is that one can interchange the expectation with the logarithm. Of course, by Jensen, one always has

$$\mathbb{E}\log Z_{N,\beta} \leq \log \mathbb{E} Z_{N,\beta}, \tag{3.4}$$

and therefore $f(\beta) \leq \beta^2/4 + \log 2$. We will indeed show later that $f(\beta) < \beta^2/4 + \log 2$ for $\beta > 1$. The proof of the above result is surprising simple and can be done by a second moment computation, proving that $\mathbb{E}Z^2 \leq \text{const} \times (\mathbf{E}Z)^2$ for $\beta < 1$. This second moment estimate is easy

$$\mathbb{E}Z^2 = \sum_{\sigma,\tau} \mathbb{E}\exp\left[\frac{\beta}{\sqrt{N}} \sum_{i<j} J_{ij} \left(\sigma_i\sigma_j + \tau_i\tau_j\right)\right]$$

$$= \exp\left[\frac{\beta^2 N}{2}\right] \sum_{\sigma,\tau} \exp\left[\frac{\beta^2}{N} \sum_{i<j} \sigma_i\sigma_j\tau_i\tau_j\right]$$

$$= (\mathbb{E}Z)^2 \left[2^{-N}\sum_\sigma \exp\left[\frac{\beta^2 N}{2}\left(\frac{1}{N}\sum_i \sigma_i\right)^2 - \frac{\beta^2}{2}\right]\right],$$

and the part in brackets is bounded for $\beta < 1$, by a simple Curie–Weiss coin tossing computation. Together with Gaussian isoperimetry (Proposition 3.1.), this proves (3.3). The original proofs in [18], and [19] were more complicated, but they derived also a much more detailed picture of the remaining fluctuations of $\log Z_N$.

There are other models like directed polymers for which one can prove that the quenched free energy equals the annealed one in certain regions, but typically, this is not possible by a simple second moment method in the full region where it is true. The fact that a second moment computation gives the result in the SK-model up to the correct critical value (for $h = 0$) is rather surprising. For $h \neq 0$, "quenched = annealed" is never true, which reveals that this is a much more interesting situation, even where β is small.

3.2 The Replica Computation

The replica trick consists in the observation that for a positive number x, one has

$$\log x = \frac{d}{dn}\exp\left(n\log x\right)\bigg|_{n=0} = \lim_{n\downarrow 0} \frac{x^n - 1}{n}.$$

If X is positive random variable, one therefore has, provided the interchange of limits is justified

$$\mathbb{E}\log X = \lim_{n\downarrow 0} \frac{\mathbb{E}X^n - 1}{n}.$$

As integer moments are often more easily evaluated then noninteger ones, the "trick" therefore is to evaluate $\mathbb{E}X^n$ for integer n, somehow extend things analytically, and performs the above limit.

For the SK-model, this is not quite the way, it is done. In fact, one just *starts* the computation of $\mathbb{E}Z_N^n$ assuming that n is an integer, but as soon as it seems convenient, one gives up this illusion and lets $n \to 0$, *before* really finishing the computation. The calculation is easy, but it is hard to swallow for a mathematician. Here it is

$$\mathbb{E}Z^n = 2^{nN}\mathbb{E}\operatorname{tr}_\sigma \exp\left[\frac{\beta}{\sqrt{N}} \sum_{1\leq i<j\leq N} J_{ij} \sum_{\alpha=1}^n \sigma_i^\alpha\sigma_j^\alpha + h\sum_{\alpha=1}^n \sum_{i=1}^N \sigma_i^\alpha\right],$$

where $\boldsymbol{\sigma} = (\sigma^\alpha)_{\alpha=1,...,n}$, $\sigma^\alpha \in \Sigma_N$, and tr_σ denotes taking averages of $\boldsymbol{\sigma}$. We interchange $\mathbb{E}$ with tr_σ, and carry out the Gauss integration over the independent J_{ij}

$$\mathbb{E}Z^n = 2^{nN} \mathrm{tr}_\sigma \exp\left[\frac{\beta^2}{2N} \sum_{1 \le i < j \le N} \left(\sum_\alpha \sigma_i^\alpha \sigma_j^\alpha\right)^2 + h\sum_{\alpha,i} \sigma_i^\alpha\right]$$

$$= 2^{nN} \mathrm{tr}_\sigma \exp\left[\frac{\beta^2}{2N} \sum_{1 \le i < j \le N} \left(n + 2\sum_{\alpha<\beta} \sigma_i^\alpha \sigma_j^\alpha \sigma_i^\beta \sigma_j^\beta\right) + h\sum_{\alpha,i} \sigma_i^\alpha\right]$$

$$= 2^{nN} \mathrm{tr}_\sigma \exp\left[\frac{\beta^2 n(N-1)}{4} + \frac{\beta^2}{2N}\left(\sum_{\alpha<\beta}\left(\sum_i \sigma_i^\alpha \sigma_i^\beta\right)^2\right.\right.$$

$$\left.\left. - \frac{n(n-1)N}{2}\right) + h\sum_{\alpha,i} \sigma_i^\alpha\right].$$

We leave out contributions in the exponent which are bounded in N. Dropping unnecessary factors, we get

$$\mathbb{E}Z^n \simeq 2^{nN} e^{\beta^2 nN/4} \mathrm{tr}_\sigma \exp\left[\frac{\beta^2}{2N} \sum_{\alpha<\beta}\left(\sum_i \sigma_i^\alpha \sigma_i^\beta\right)^2 + h\sum_{\alpha,i} \sigma_i^\alpha\right].$$

Now, we do not like the square in the exponent, and linearize it with Gaussian integrals. Let therefore $g_{\alpha\beta}$, $\alpha < \beta$, be standard Gaussians, whose expectations are denoted by E, and we get

$$\mathbb{E}Z^n \simeq 2^{nN} e^{\beta^2 nN/4} E \, \mathrm{tr}_\sigma \exp\left[h\sum_\alpha \sum_{i=1}^N \sigma_i^\alpha\right] \exp\left[\frac{\beta}{\sqrt{N}} \sum_{\alpha<\beta} g_{\alpha\beta} \sum_i \sigma_i^\alpha \sigma_i^\beta\right].$$

Now, we happily can perform the tr_σ operation, individually on each spin components, and each i gives the same contribution. Therefore

$$\mathbb{E}Z^n \simeq 2^{nN} e^{\beta^2 nN/4} E \left\{\mathrm{tr}_\sigma \exp\left[h\sum_\alpha \sigma^\alpha + \frac{\beta}{\sqrt{N}} \sum_{\alpha<\beta} g_{\alpha\beta} \sigma^\alpha \sigma^\beta\right]\right\}^N,$$

where now $\boldsymbol{\sigma} = (\sigma^\alpha)$ has only one component, i.e., $\sigma^\alpha \in \{-1, 1\}$. We write out the Gaussian integrals as $\int dq_{\alpha\beta} \exp\left[-q_{\alpha\beta}^2/2\right]$, the prefactor being of no importance for us, but we make a linear variable transformation replacing $q_{\alpha\beta}$ by $q_{\alpha\beta}/\beta\sqrt{N}$. The coefficient from the variable transformation is again of no importance, and we get

$$\mathbb{E}Z^n \simeq 2^{nN} \int d\mathbf{q} \exp\left[-\frac{N\beta^2}{2} \sum_{\alpha<\beta} q_{\alpha\beta}^2 + N \log \mathrm{tr}_\sigma \, e^{L(\mathbf{q},\sigma)} + \frac{\beta^2 nN}{4}\right]$$

$$= 2^{nN} \int d\mathbf{q} \exp\left[nN\left\{-\frac{\beta^2}{2n} \sum_{\alpha<\beta} q_{\alpha\beta}^2 + \frac{1}{n} \log \mathrm{tr}_\sigma \, e^{L(\mathbf{q},\sigma)} + \frac{\beta^2}{4}\right\}\right],$$

where
$$L(\mathbf{q}, \boldsymbol{\sigma}) \overset{\text{def}}{=} \beta^2 \sum_{\alpha < \beta} q_{\alpha\beta} \sigma^\alpha \sigma^\beta + h \sum_\alpha \sigma^\alpha.$$

Now, one evaluates the integral by Laplace approximation. Naively, one thinks that one should take the maximum over the $\{\cdot\}$, but just in time it comes back that we are not really interested in the integer case and that $n \to 0$. As the number of summands in $\sum_{\alpha < \beta}$ is $n(n-1)/2$, and therefore negative in the region of interest, the feeling is that one should take the *minimum* instead. Therefore

$$\lim_{N \to \infty} \frac{1}{N} \mathbb{E} \log Z_N = \lim_{N \to \infty} \frac{1}{N} \lim_{n \to 0} \frac{\mathbb{E} Z_N^n - 1}{n}$$

$$= \inf_{\mathbf{q}} \lim_{n \to 0} \left\{ -\frac{\beta^2}{2n} \sum_{\alpha < \beta} q_{\alpha\beta}^2 + \frac{1}{n} \log \operatorname{tr}_{\boldsymbol{\sigma}} e^{L(\mathbf{q}, \boldsymbol{\sigma})} + \frac{\beta^2}{4} \right\} + \log 2.$$

$$(3.5)$$

There is also an issue of interchanging the $N \to \infty$ and $n \to 0$ limit. Even in the physics literature this interchange is considered as somewhat "problematic," but the final formula looks interesting anyway, if one is not too picky about the fact that it lacks a decent mathematical meaning. The showdown, however, just starts now, namely to find the infimum. Sherrington–Kirkpatrick made short work of the problem and assumed that there is no sufficient reason why the n replicas should behave "asymmetric," and put $q_{\alpha\beta} = q$. Assuming this, we have that $\sum_{\alpha < \beta} q_{\alpha\beta}^2 = n(n-1)q^2/2$, and an n cancels out. Furthermore, the $n \to 0$ limit in this part is no longer particularly demanding: $\lim_{n \to 0}(n-1) = -1$. Taking the $n \to 0$ limit in the other part is only a bit more tricky. In the replica-symmetric case, we have

$$L(\mathbf{q}, \boldsymbol{\sigma}) = \frac{\beta^2 q}{2} \left(\left(\sum_\alpha \sigma^\alpha \right)^2 - n \right) + h \sum_\alpha \sigma^\alpha,$$

$$e^{L(\mathbf{q}, \boldsymbol{\sigma})} = e^{-q\beta^2 n/2} \exp\left[h \sum_\alpha \sigma^\alpha \right] E \exp\left[g\beta\sqrt{q} \sum_\alpha \sigma^\alpha \right]$$

$$\operatorname{tr}_{\boldsymbol{\sigma}} e^{L(\mathbf{q}, \boldsymbol{\sigma})} = e^{-q\beta^2 n/2} E \left[\cosh(g\beta\sqrt{q} + h) \right]^n,$$

$$\frac{1}{n} \log \operatorname{tr}_{\boldsymbol{\sigma}} e^{L(\mathbf{q}, \boldsymbol{\sigma})} = -q\beta^2/2 + \frac{1}{n} \log E \left[\cosh(g\beta\sqrt{q} + h) \right]^n$$

$$= -q\beta^2/2 + \frac{1}{n} \log E \exp\left[n \log \cosh(g\beta\sqrt{q} + h) \right]$$

$$\simeq -q\beta^2/2 + \frac{1}{n} \log E \left(1 + n \log \cosh(g\beta\sqrt{q} + h) \right), \quad n \simeq 0$$

$$\to -q\beta^2/2 + E \log \cosh(g\beta\sqrt{q} + h).$$

Summing things together, we get

$$\lim_{N\to\infty} \frac{1}{N} \mathbb{E} \log Z_N = RS(\beta, h)$$

$$\stackrel{\text{def}}{=} \inf_q \left\{ \frac{\beta^2}{4} (1-q)^2 + \int \log \cosh (x\beta\sqrt{q} + h) \frac{1}{\sqrt{2\pi}} e^{-x^2/2} dx \right\}.$$

$$(3.6)$$

This is the replica-symmetric "solution" of the SK-model.

For later use, we try to find the minimizing q. Differentiating with respect to q, and setting it 0 gives the equation

$$\beta (1-q) = \frac{1}{\sqrt{q}} \int \tanh (h + \beta\sqrt{q}x) \, x \frac{1}{\sqrt{2\pi}} e^{-x^2/2} dx,$$

and using partial integration, and $(\tanh)' = 1 - \tanh^2$, we get

$$\beta (1-q) = \beta \int \left[1 - \tanh^2 (h + \beta\sqrt{q}x) \right] \frac{1}{\sqrt{2\pi}} e^{-x^2/2} dx,$$

i.e.,

$$q = \int \tanh^2 (h + \beta\sqrt{q}x) \frac{1}{\sqrt{2\pi}} e^{-x^2/2} dx. \qquad (3.7)$$

For $h = 0$, $q = 0$ is always a solution, and for $\beta \leq 1$, it is the only one, as one can easily check. Therefore RS $(\beta, 0) = \beta^2/4$ for $\beta \leq 1$. For $\beta > 1$, there is, however, at least one other solution of this fixed point equation, which follows easily by calculating the derivative of the expression on the right-hand side at $q = 0$. In fact, there is just *one* other solution $q(\beta) > 0$ which gives the minimum, and therefore RS $(\beta, 0) < \beta^2/4$ for $\beta > 1$.

For $h > 0$, (3.7) does have a unique positive solution.

Lemma 3.3. *Let $\beta, h > 0$ be arbitrary. Then (3.7) has a unique solution $q(\beta, h)$.*

I am not going to prove this here. The proof is due to Guerra and is short but a bit tricky. Talagrand has it in his book ([3], Proposition 2.4.8).

3.3 The Parisi Ansatz

The main question is whether $f(\beta, h) = RS(\beta, h)$. It is certainly correct for $h = 0$ and $\beta \leq 1$, as we have seen before. However, for $\beta > 1$, it is not correct. This is far from trivial to see. It will, however, turn out that for $h \neq 0$, the formula is correct again for small β, but not for large ones. Even the small β case is highly nontrivial. That the solution cannot be correct for large β was already realized by Sherrington and Kirkpatrick by calculating the entropy, which has to be positive, but it can also be computed from the free energy, and if one uses RS, it becomes negative for large β. So already Sherrington and Kirkpatrick concluded that their own solution is not correct for large β.

The RS solution is supposed to be correct for β up to the celebrated AT line (de Almeida–Thouless line [20]), i.e., for β satisfying

$$\beta^2 \int \frac{1}{\cosh^4\left(h + \beta\sqrt{q\,(\beta,h)}x\right)} \frac{1}{\sqrt{2\pi}} e^{-x^2/2} dx < 1, \tag{3.8}$$

but this is not yet proved, but it is now simply a nasty analytical problem, as the Parisi formula for $f(\beta, h)$ is proved for the whole temperature region. (The above condition comes up through a local stability computation.)

In order to overcome the problem with the replica-symmetric solution for large β, there had been various proposals for a different ansatz for the minimizing problem in (3.5), no longer assuming that all the $q_{\alpha\beta}$ are equal. This is the famous "replica symmetry breaking." A particular ansatz for this is due to Parisi. The ansatz makes a *very special* assumption on the matrix $Q = (q_{\alpha\beta})$, namely that it has a kind of hierarchical organization. The question then remained if there could not be a better choice not satisfying the Parisi ansatz. A justification of the Parisi ansatz before Talagrand's proof was the proof that it is in a sense locally stable, by computing Hessians, and that it was the only one found having this property, but the really convincing argument was that the outcome had interesting consequences also outside the "replica formulation." Very nice explanations of these issues can be found in [21]. Here just a cursory explanation of what is going on.

The replica-symmetric ansatz fixes the matrix Q to be of the following form:

$$Q = \begin{pmatrix} 0 & q & q & \cdots & \cdots & q \\ & 0 & q & \cdots & \cdots & q \\ & & 0 & q & \cdots & q \\ & & & \ddots & & \vdots \\ & & & & 0 & q \\ & & & & & 0 \end{pmatrix}.$$

In the Parisi ansatz, one uses more complicated matrices. There are a number of levels. In the end, this number has to go to infinity, but let us first look at the simplest case, the case with one level of replica symmetry breaking. Here one takes a matrix of the form

$$\left(\begin{pmatrix} 0 & q_2 & q_2 \\ & 0 & q_2 \\ & & 0 \end{pmatrix} \begin{pmatrix} q_1 & q_1 & q_1 \\ q_1 & q_1 & q_1 \\ q_1 & q_1 & q_1 \end{pmatrix} \\ \begin{pmatrix} 0 & q_2 & q_2 \\ & 0 & q_2 \\ & & 0 \end{pmatrix} \right).$$

The rule is that one divides the $n \times n$ matrix by choosing $n_1 \leq n$ such that n/n_1 is an integer, and then one divides the matrix into $(n/n_1)^2$

submatrices of the form $n_1 \times n_1$. The diagonal blocks get q_2 above the diagonal, and the off-diagonal blocks all get q_1. In the above example, one has $n = 6$ and $n_1 = 3$. Then one does the computation analogously as above, keeps $m_1 \stackrel{\text{def}}{=} n_1/n$ fixed, and lets formally $n \to 0$. This leads to a variational problem. One can check that one can always assume that $0 \le q_1 \le q_2 \le 1$. For β small it turns out that nothing new is achieved. The optimal choice for the qs is $q_1 = q_2$, but for large β, some $q_1 < q_2$ give a lower value. This is the "one level symmetry breaking," but one can proceed by dividing the q_2 blocks in a similar fashion, which leads to a "two level symmetry breaking," and one can go on in this way with an arbitrary number of symmetry breakings. The calculations are somewhat lengthy but not difficult. Here is the outcome.

Let $K \in \mathbb{N}$ (the number of symmetry breakings), and then we choose parameters

$$0 = m_0 < m_1 < \ldots < m_{K-1} < m_K = 1, \tag{3.9}$$

$$0 = q_0 \le q_1 < \ldots < q_K < q_{K+1} = 1. \tag{3.10}$$

For $i = 0, \ldots, K$ let g_i be Gaussian with variance $\beta^2 (q_{i+1} - q_i)$, and set $Y_{K+1} \stackrel{\text{def}}{=} \cosh\left(h + \sum_{i=0}^{K} g_i\right)$. Then one defines

$$Y_K \stackrel{\text{def}}{=} \left[E_K \left(Y_{K+1}^{m_K}\right)\right]^{1/m_K} = E_K \left(Y_{K+1}\right). \tag{3.11}$$

where E_K means that one integrates out g_K, so that Y_K still depends on $g_0, \ldots, g_{K-1}$. Then one defines

$$Y_{K-1} \stackrel{\text{def}}{=} \left[E_{K-1} \left(Y_K^{m_{K-1}}\right)\right]^{1/m_{K-1}}$$

and so on, until one gets Y_1. Y_1 is still a random variable as it depends on g_0. Remark, however, that in case $q_1 = 0$ which we do not exclude, there is no randomness left. In any case, we set

$$\mathcal{P}_K (m, q; \beta, h) \stackrel{\text{def}}{=} E \log Y_1 - \frac{\beta^2}{4} \sum_{i=1}^{K} m_i \left(q_{i+1}^2 - q_i^2\right) + \log 2. \tag{3.12}$$

Then $\inf_{m,q} \mathcal{P}_K (m, q)$ is the value one obtains by optimizing (3.5) with the Parisi ansatz at K levels of replica symmetry breaking, and therefore, believing that first of all the replica trick works, and secondly that the ansatz of Parisi finds the minimum, we get

$$f(\beta, h) = \inf_{K, m, q} \mathcal{P}_K (m, q) = \lim_{K \to \infty} \inf_{m,q} \mathcal{P}_K (m, q). \tag{3.13}$$

Theorem 3.4. (Parisi Formula) *The Parisi formula (3.13) is correct for all β, h.*

The proof is due to Guerra [5] who proved the upper bound, and to Talagrand [6] who then finished the proof.

In the case of the SK-model, either one has $K = 1$, which gives the true value in the region where the replica-symmetric solution is correct, or one has to take $K \to \infty$, and therefore one has "replica symmetry breaking" at infinitely many levels. There are other models, with the minimum assumed at one level of symmetry breaking, i.e., $K = 2$. One can artificially cook up cases with arbitrary K, but $K = 1, 2, \infty$ seem to be the only ones coming up "naturally." In the case $K = \infty$, one can phrase the limit $K \to \infty$ directly as a variational problem involving continuous functions $q \to x\,(q)$. The finite K case then corresponds to taking step functions $x\,(q) = m_i$ for $q \in [q_i, q_{i+1})$.

Here is an outline of what the physicists believe to be the picture behind the RS solution $(K = 1)$, and the replica symmetry breaking $(K > 1)$. This picture emerged partly from another nonrigorous approach, the so-called "cavity method" which led to the same formula for the free energy, and gave a clearer picture about the Gibbs distribution (see [2]).

The region, where the RS solution (3.6) is valid, is characterized by the property that the σ_i under the Gibbs measure are still "fairly independent." The $h = 0$ case is simple because, due to symmetry, the expectation under the Gibbs measure is 0. For $h \neq 0$, the expectation of σ_i under the Gibbs measure $\mathcal{G}_{N,\beta,h,\omega}$ is $m_i \stackrel{\mathrm{def}}{=} \mathcal{G}\,(\sigma_i)$ which satisfies $\mathbb{E} m_i^2 = q\,(\beta, h)$, q being the solution of (3.7), equality in the $N \to \infty$ limit. The m_i are themselves approximately independent under the measure $\mathbb{P}$. One therefore has the following picture (for large N): the randomness of the disorder (i.e., the J_{ij}) produce the nearly i.i.d. random variables m_i, and given the disorder, the Gibbs measure has approximately independent spin variables σ_i with mean m_i. The property that the σ_i are approximately independent is reflected in the physics community saying that there is just "one pure state."

Given this picture, $q\,(\beta, h)$ has a precise mathematical interpretation in terms of the Gibbs measure. It is the almost sure limit (as $N \to \infty$) of the overlaps of two independent realizations of the spin variables

$$R_N\,(\sigma, \sigma') = \frac{1}{N} \sum_{i=1}^{N} \sigma_i \sigma_i' \simeq \frac{1}{N} \sum_{i=1}^{N} m_i^2 \simeq q\,(\beta, h) \qquad (3.14)$$

by the law of large numbers. The precise statement is as follows: Let $\nu_N^{(2)}$ be the measure on $\Sigma_N \times \Sigma_N$ defined by

$$\nu_N^{(2)}\,(\sigma, \sigma') \stackrel{\mathrm{def}}{=} \int \mathbb{P}\,(d\omega)\, \mathcal{G}_{N,\omega}^{\otimes 2}\,(\sigma, \sigma'), \qquad (3.15)$$

where $\mathcal{G}^{\otimes 2}$ denotes the twofold product Gibbs measure. Then for small enough β

$$\lim_{N \to \infty} \nu_N^{(2)}\,(|R_N\,(\sigma, \sigma') - q\,(\beta, h)| \geq \varepsilon) = 0, \quad \forall \varepsilon > 0.$$

This means that the overlap of independent replicas is self-averaging. The high-temperature regime is now mathematically very well understood, mainly through the work of Talagrand (see [3], Chap. 2).

In the low-temperature regime things become much more complicated. First of all, the RS solution is no longer correct, but this is only one aspect. The overlaps are no longer self-averaging but stay random. The Gibbs distribution splits into a "countable number of pure states," a statement made in the physics literature which is difficult to make mathematically precise. Essentially the "pure states" under the Gibbs distribution should be organized in a hierarchical way. This hierarchy somehow reflects the hierarchical ansatz in the Parisi matrices above. Nothing of this has been proved mathematically, and probably not all statements made in the physics literature should be taken (mathematically) too literally. See also the discussion in Sect. 6.

4 Mathematically Rigorous Results for the SK-Model

4.1 Guerra's Interpolation Method

One of the breakthroughs in a mathematical understanding of the SK-model was the idea of Guerra to use what in probability theory is called the "smart path" method. It consists in interpolating between the model one is interested in, and a much simpler one, in such a way, that one has some control over the derivative along the path joining the two models. The trick is of course to choose the interpolating path in a clever way (and also the simple model).

I will explain this in the simplest case, where one proves that the replica-symmetric solution is a strict bound for the free energy, for all N and in the full region of parameters. I first learnt about this argument at a conference in Vulcano in September 1998 where Francesco Guerra explained it in his talk. At that time, nobody (perhaps except Guerra) realized how important this argument would become.

As a preparation, we differentiate $f_N(\beta, h)$ with respect to β

$$\frac{\partial f_N(\beta, h)}{\partial \beta} = \frac{1}{N^{3/2}} \sum_{i<j} \sum_{\sigma} \sigma_i \sigma_j \mathbb{E}\left(\frac{J_{ij}}{Z_N(\beta, h)} \exp\left[-H_{N,\beta,h}(\sigma)\right]\right).$$

We use Wick's theorem to get rid of the J_{ij}

$$\mathbb{E}\left(\frac{J_{ij}}{Z_N(\beta, h)} \exp\left[-H_{N,\beta,h}(\sigma)\right]\right)$$

$$= \mathbb{E}\left(\left(\frac{\partial}{\partial J_{ij}} \frac{1}{Z_N(\beta, h)}\right) \exp\left[-H_{N,\beta,h}(\sigma)\right]\right)$$

$$+ \mathbb{E}\left(\frac{1}{Z_N(\beta, h)} \frac{\partial}{\partial J_{ij}} \exp\left[-H_{N,\beta,h}(\sigma)\right]\right)$$

$$= -\mathbb{E}\left(\sum_{\tau} \frac{\beta}{\sqrt{N}} \tau_i \tau_j \mathcal{G}(\tau) \mathcal{G}(\sigma)\right) + \mathbb{E}\left(\frac{\beta}{\sqrt{N}} \sigma_i \sigma_j \mathcal{G}(\sigma)\right).$$

Therefore, we get

$$\frac{\partial f_N(\beta, h)}{\partial \beta} = \frac{\beta}{2} \left(1 - \sum_\sigma \sum_\tau R_N(\sigma, \tau)^2 \, \mathbb{E} \left(\mathcal{G}(\tau) \mathcal{G}(\sigma) \right) \right)$$

$$= \frac{\beta}{2} \left(1 - \nu_N^{(2)} \left(R_N(\sigma, \tau)^2 \right) \right), \qquad (4.1)$$

where $\nu_N^{(m)}$ is the measure on Σ_N^m obtained from taking the m-fold product measure of the Gibbs distribution, and integrate it over $\mathbb{P}$ (see (3.16)).

Theorem 4.1. *For all $\beta > 0$, $h \in \mathbb{R}$, and $N \in \mathbb{N}$ one has*

$$f_N(\beta, h) \leq \mathrm{RS}(\beta, h),$$

where $\mathrm{RS}(\beta, h)$ is defined by (3.6).

Proof. The proof is by interpolation. Let for an arbitrary number $q \geq 0$, and $t \in [0, 1]$

$$X(t, \sigma) \overset{\text{def}}{=} \sqrt{\frac{t}{N}} \sum_{1 \leq i < j \leq N} J_{ij} \sigma_i \sigma_j + \sqrt{1 - t} \sum_{i=1}^N \sqrt{q} g_i \sigma_i,$$

$$-H_{\beta, h}(t, \sigma) \overset{\text{def}}{=} \beta X(t, \sigma) + h \sum_{i=1}^N \sigma_i, \qquad (4.2)$$

where g_i is a set of standard Gaussian variables, independent of the Js. The basic idea of this interpolation is to relate the Hamiltonian we are interested in with a much simpler one with independent σ_i, which, however, have the right overlap structure. For $t = 0$, we have, conditioned on the g_i, independent spins with mean $\tanh \left(h + \beta \sqrt{q} g_i \right)$. Therefore, if we take two independent realizations σ, σ' (still conditioned on the g_i), we get

$$\frac{1}{N} \sum_{i=1}^N \sigma_i \sigma_i' \simeq \frac{1}{N} \sum_{i=1}^N \tanh^2 \left(h + \beta \sqrt{q} g_i \right)$$

$$\simeq \int \tanh^2 \left(h + \beta \sqrt{q} x \right) \frac{1}{\sqrt{2\pi}} e^{-x^2/2} dx$$

which equals q, if we take for q the solution of (3.7). The clever idea by Guerra is that one can control what happens along the path from $t = 0$ to 1. For the moment, we have not even to assume that q is the right one, and we can just take it arbitrary ≥ 0. We again define the partition function

$$\xi_{\beta, h}(t) \overset{\text{def}}{=} \sum_\sigma \exp \left[-H_{\beta, h}(t, \sigma) \right],$$

and we write $\mathcal{G}_{\beta,h}(t,\sigma)$ for the corresponding Gibbs measure. Let

$$\phi(t) \stackrel{\text{def}}{=} \frac{1}{N}\mathbb{E}\log\xi(t). \tag{4.3}$$

Remark that

$$\phi(0) = \int \log\cosh(\beta\sqrt{q}x + h)\frac{1}{\sqrt{2\pi}}e^{-x^2/2}dx + \log 2, \tag{4.4}$$

$$\phi(1) = f_N(\beta, h).$$

We compute the derivative of $\phi(t)$ with respect to t. For the derivative with respect to the first occurrence, we can use (4.1), replacing β by $\beta\sqrt{t}$, and multiply it with $\beta/2\sqrt{t}$. So we get

$$\phi'(t) = \frac{\beta^2}{4}\left(1 - \nu_N^{(2)}\left(R_N(\sigma,\tau)^2\right)\right) - \frac{\sqrt{q}}{2\sqrt{1-t}}\sum_\sigma \sigma_i \sum_i \mathbb{E}g_i \frac{\exp[\Phi(t,\sigma)]}{\xi(t)}.$$

Similarly, by a computation leading to (4.1), we get for the second summand

$$(\beta^2 q/2)\left(1 - \nu_{N,t}^{(2)}(R_N(\sigma,\tau))\right),$$

and therefore

$$\begin{aligned}
\phi'(t) &= \frac{\beta^2}{4}\nu_{N,t}^{(2)}\left(1 - R_N(\sigma,\tau)^2 - 2q\left(1 - R_N(\sigma,\tau)\right)\right)\\
&= \frac{\beta^2}{4}\left\{(1-q)^2 - \nu_{N,t}^{(2)}\left((R_N(\sigma,\tau) - q)^2\right)\right\}\\
&\le \frac{\beta^2}{4}(1-q)^2.
\end{aligned}$$

$$\phi(1) - \phi(0) \le \frac{\beta^2}{4}(1-q)^2.$$

We therefore get from (4.4) for *any* N and *any* $q \ge 0$

$$f_N(\beta, h) \le \frac{\beta^2}{4}(1-q)^2 + \int \log\cosh(h + \beta\sqrt{q}x)\frac{1}{\sqrt{2\pi}}e^{-x^2/2}dx + \log 2.$$

Taking the infimum over q of the right-hand side implies the theorem.   □

A corollary of Guerra's bound is that $f(\beta, h) < \beta^2/4 + \log 2$ for $\beta > 1$, simply because $\text{RS}(\beta, 0) < \beta^2/4 + \log 2$ for $\beta > 1$, as is easily checked. This fact was first proved by Comets [22] using different methods.

A very important point is that the proof does not only give the desired result, but also gives an expression of the difference, namely

$$\text{RS}(\beta, h) - f_N(\beta, h) = \frac{\beta^2}{4}\int_0^1 \nu_{N,t}^{(2)}\left((R_N(\sigma,\tau) - q)^2\right)dt. \tag{4.5}$$

In order to prove that $\mathrm{RS}\,(\beta, h) = f_N\,(\beta, h)$, one therefore "only" has to show that for the optimal q (i.e., the one given by (3.7)), one has $R_N\,(\sigma, \tau) \simeq q$ with large $\nu_{N,t}^{(2)}$ probability, at least in the t-average. This is not true for large β, but it is true for small β.

It should also be emphasized that the Gibbs measure $\mathcal{G}_t$ structurally is not much different from the original measure. In fact it is of the form

$$\frac{1}{Z} \exp\left[\beta' H_N\,(\sigma) + \gamma \sum_i g_i \sigma_i + h \sum_i \sigma_i\right],$$

where the g_i are new independent Gaussians, and γ is an additional parameter.

The proof that the replica-symmetric solution is the correct one for small β (also including $h \neq 0$) was first given by Talagrand using the cavity method discussed in the next Sect. 4.2, but it can also be proved by a refinement of Guerra's method. What one essentially does is to copy the interpolation method for a replicated system, which leads to bounds on the replicated system with which one can prove that the right-hand side of (4.5) goes to 0 if q is chosen properly. This method is explained in Talagrand's book. A particularly nice and simple version is due to Guerra and Toninelli [23] who worked with a replicated system where the replicas are coupled quadratically. Their interpolated and replicated Hamiltonian of two spin configurations σ, σ' has two parameters $t \in [0, 1]$, and $\lambda > 0$

$$X^{(2)}\,(t, \sigma, \sigma') \stackrel{\mathrm{def}}{=} \sqrt{\frac{t}{N}} \sum_{1 \leq i < j \leq N} J_{ij}\,(\sigma_i \sigma_j + \sigma_i' \sigma_j') + \sqrt{1-t} \sum_{i=1}^{N} \sqrt{q} g_i\,(\sigma_i + \sigma_i'),$$

$$- H^{(2)}\,(t, \lambda, \sigma, \sigma', \beta, h) \stackrel{\mathrm{def}}{=} \beta X^{(2)}\,(t, \sigma, \sigma') + h \sum_i (\sigma_i + \sigma_i') + \frac{\lambda}{2} N \left(R_N\,(\sigma, \sigma') - q\right)^2.$$

The problem of choosing the right q is showing up already at $t = 0$, where the interaction among the different spin sites is absent. To see this, consider spins σ_i which are just distributed with a Hamiltonian

$$-H\,(\sigma) = \sum_i \beta \sqrt{q} g_i \sigma_i + h \sum_i \sigma_i.$$

For fixed g_i, they are independent with mean $\tanh\left(h + \beta \sqrt{q} g_i\right)$. Therefore, for an independent replica σ' of such a spin family, one gets

$$R_N\,(\sigma, \sigma') \simeq \frac{1}{N} \sum_{i=1}^{N} \tanh^2 (h + \beta \sqrt{q} g_i) \simeq \int \tanh^2 (h + \beta \sqrt{q} x) \frac{1}{\sqrt{2\pi}} e^{-x^2/2} dx.$$

Therefore, if q satisfies (3.7), one has, under the replicated Gibbs distribution at $t = 0$, that $R_N\,(\sigma, \sigma') \simeq q$. It is then not difficult to show essentially by Curie–Weiss type computations that

$$\Delta_N\left(\beta,h,\lambda\right) \stackrel{\text{def}}{=} \frac{1}{N}\mathbb{E}\log\mathcal{G}_{t=0}^{\otimes 2}\left(\exp\left[\frac{\lambda}{2}N\left(R_N\left(\sigma,\sigma'\right)-q\right)^2\right]\right)\simeq 0,$$

if λ is not too large. By a clever interpolation scheme involving t, and λ depending on t, Guerra and Toninelli then relate $\nu_{N,t}^{(2)}\left(\left(R_N\left(\sigma,\tau\right)-q\right)^2\right)$ to $\Delta_N\left(\beta,h,\lambda\right)$, and show that for small enough β, this is close to 0, along the whole path from $t=0$ to 1, which proves $f\left(\beta,h\right)=\text{RS}\left(\beta,h\right)$. For details, see [23].

In Talagrand's version, explained in his book, he investigates the coupled system with fixed overlap $R_N\left(\sigma,\sigma'\right)=u$, and derives Guerra-type bounds with which he is able to discuss conditions under which $\nu_{N,t}^{(2)}\left(\left(R_N\left(\sigma,\tau\right)-q\right)^2\right)$ $\simeq 0$. His technique played a crucial rôle in the derivation of the Parisi formula (Theorem 3.4.). See his article in this volume [12].

4.2 Mathematical Variants of the Cavity Method

The basic idea of the so-called the "cavity method" is to investigate the influence of a new additional spin variable on a system with N-spins. We consider the standard SK-Hamiltonian, but now with $N+1$. We then write the Hamiltonian in terms of the Hamiltonian on N-spin variables $\sigma_1,\ldots,\sigma_N$ acting on the "newcomer" $\tau=\sigma_{N+1}$

$$\frac{\beta}{\sqrt{N+1}}\sum_{1\leq i<j\leq N+1}J_{ij}\sigma_i\sigma_j+h\sum_{i=1}^{N+1}\sigma_i$$

$$=\sqrt{\frac{N}{N+1}}\frac{\beta}{\sqrt{N}}\sum_{1\leq i<j\leq N}J_{ij}\sigma_i\sigma_j+h\sum_{i=1}^{N}\sigma_i+\frac{\beta}{\sqrt{N+1}}\left(\sum_{i=1}^{N}J_{i,N+1}\sigma_i\right)\tau+h\tau.$$

$$(4.6)$$

We can replace $\sqrt{N+1}$ by $\sqrt{N}$ in the third summand, as the error when doing so is only of order $O\left(N^{-1/2}\right)$ and can be neglected in the $N\to\infty$ limit. Replacing the $\sqrt{N/(N+1)}$ in the first factor by 1 requires, however, a correction by a summand which is stochastically of order 1. If we set $\beta'\stackrel{\text{def}}{=}\sqrt{N/(N+1)}\beta$ and define the cavity variables

$$y_\sigma\stackrel{\text{def}}{=}\frac{1}{\sqrt{N}}\sum_{i=1}^{N}J_{i,N+1}\sigma_i,$$

we have

$$-H_{N+1,\beta,h}\left(\sigma,\tau\right)\simeq -H_{N,\beta',h}\left(\sigma\right)+\beta y_\sigma\tau+h\tau.$$

Remark that the cavity variables y_σ are independent of the N-Hamiltonian $H_{N,\beta',h}$, and in fact the y_σ are just Gaussian random variables with covariances

$$\mathbb{E} y_\sigma y_{\sigma'} = R_N (\sigma, \sigma').$$

The basic idea of the cavity method in the physics literature is that for large N, the newcomers should leave the distribution of the Gibbs structure invariant, and that the distribution of the overlap of "replicated newcomers" should correspond to the distribution of the overlaps in the N-system. It is difficult to give this a precise mathematical meaning. What is done, is to "assume" that large systems are hierarchically organized, and one tries to characterize this hierarchical structure by a self-consistency property. Although some of the main predictions of the Parisi theory have been verified by now, a proof that the SK-model is asymptotically hierarchically structured is still completely lacking. Mathematically, one aspect of this would be the validity of the so-called "ultra metricity" conjecture (see Sect. 6). The abstract mathematical structure of the self-consistency has been worked out in [24], but there is no proof there that the SK-model has asymptotically this structure. I believe that what is lacking is a proper concept of a "contraction toward ultrametricity," and a proof that adding spins to a large system performs such a contraction. The hierarchical structure which is believed (and in some cases proved) to show up is that of the generalized random energy model in Ruelle's formulation, which I will shortly discuss in Sect. 5.2. This is still very much on a speculative level, but such concepts play a major rôle in Talarand's versions of the cavity method which he applied to many models. In his first paper on the SK-model, he proved such a contraction property for the small β case, proving $f(\beta, h) = \mathrm{RS}(\beta, h)$ for small enough β. I give an outline of this argument. Although, for the SK-model, it is now partly outdated by the use of Guerra's interpolation, there are many models where there is no variant yet of Guerra's estimates, and where Talagrand's cavity method is the only one available.

Lemma 4.2. *Let* $\gamma_N (\beta, h) \stackrel{\text{def}}{=} \mathbb{E} \left(\mathrm{cov}_{\mathcal{G}} (\sigma_1, \sigma_2)^2 \right)$. *Then, for small enough* $\beta > 0$ *there exists* $\rho < 1$ *with*

$$\gamma_{N+2} \left(\beta \sqrt{1 + 2/N}, h \right) \le \rho \gamma_N (\beta, h) + O \left(N^{-1} \right).$$

I am not giving the proof here. It is based on adding two newcomers to the N-system. The direct interaction of the newcomers can be neglected, and their correlation in the $(N + 2)$ system (at slightly changed temperature parameter) can be expressed entirely through the influence of the N-system on the newcomers. After some computations, one obtains the above contraction property for small β. The conclusion from this property is that for $i \ne j$, the variables σ_i, σ_j become approximately independent under the Gibbs measure. Let

$$m_i \stackrel{\text{def}}{=} \mathbb{E}_{\mathcal{G}} \sigma_i.$$

These are random variables because they depend on the random environment.

An important consequence of $\gamma_N \simeq 0$ is the following one. Let $(g_i)_{1 \le i \le N}$ be a sequence of independent Gaussians, independent also of the interaction variables J_{ij}. Then for $\alpha > 0$, with overwhelming probability (with respect to the g_i and J_{ij}), one has

$$
E_{\mathcal{G}} \exp \left[\frac{\alpha}{\sqrt{N}} \sum_{i=1}^{N} g_i \sigma_i \right]
$$

$$
\simeq \exp \left[\frac{\alpha}{\sqrt{N}} \sum_{i=1}^{N} g_i m_i \right] E_{\mathcal{G}} \exp \left[\frac{\alpha^2}{2N} \sum_{i=1}^{N} (\sigma_i - m_i)^2 \right]. \tag{4.7}
$$

This follows by a simple application of the Cauchy–Schwarz inequality. The basic point for the validity of $f(\beta, h) = \mathrm{RS}(\beta, h)$ for small β is that the m_i become asymptotically (roughly) independent for large N (when β is small, this is false for large β). I give a quick outline of the argument that near independence of the m_i implies the validity of the replica-symmetric solution. The starting point is a simple computation of the derivatives with respect to β

$$
\frac{\partial \mathrm{RS}(\beta, h)}{\partial \beta} = \frac{\beta}{2} \left(1 - q(\beta, h)^2 \right),
$$

$$
\frac{\partial f_N(\beta, h)}{\partial \beta} = \frac{\beta}{2} \left(1 - \mathbb{E} \frac{1}{N^2} \sum_{i,j} E_{\mathcal{G}} (\sigma_i \sigma_j)^2 \right).
$$

The first equation is by a simple computation, and the second is (4.1). Given the property that the σ_i are approximately independent under the Gibbs distribution, one has

$$
\frac{1}{N^2} \sum_{i,j} E_{\mathcal{G}} (\sigma_i \sigma_j)^2 \simeq \frac{1}{N^2} \sum_{i,j} m_i^2 m_j^2 = \left(\frac{1}{N} \sum_i m_i^2 \right)^2.
$$

If the the m_i are (approximately) independent under $\mathbb{P}$, then one would get

$$
\mathbb{E} \frac{1}{N^2} \sum_{i,j} E_{\mathcal{G}} (\sigma_i \sigma_j)^2 = \left(\mathbb{E} m_1^2 \right)^2.
$$

It is not difficult to see that in such a case $\mathbb{E} m_1^2$ would, in the $N \to \infty$ limit, have to match $q(\beta, h)$. Here is a sketch of the argument. By (4.2), one gets

$$
\mathbb{E} \phi_{N+1} \left(\beta \sqrt{\frac{N+1}{N}} \right) = \mathbb{E} m_{N+1}^2 \left(\beta \sqrt{\frac{N+1}{N}} \right) = \mathbb{E} \left(\frac{E_{\mathcal{G}_{N,\beta}} \sinh(\beta y_\sigma + h)}{E_{\mathcal{G}_{N,\beta}} \cosh(\beta y_\sigma + h)} \right)^2,
$$

where

$$
\phi_N(\beta) \overset{\text{def}}{=} \frac{1}{N} \sum_i m_i^2.
$$

If $\gamma_N (\beta, h) \simeq 0$, then from (4.7) one gets

$$\frac{E_{\mathcal{G}_{N,\beta}} \sinh (\beta y_\sigma + h)}{E_{\mathcal{G}_{N,\beta}} \cosh (\beta y_\sigma + h)} \simeq \tanh \left(\beta E_{\mathcal{G}_{N,\beta}} y_\sigma + h \right) = \tanh \left(\frac{\beta}{\sqrt{N}} \sum_{i=1}^{N} J_{i,N+1} m_i + h \right).$$

If the m_i are approximately independent under $\mathbb{P}$, then ϕ_N is essentially a constant, and we would get

$$\phi_{N+1} \left(\beta \sqrt{\frac{N+1}{N}} \right) \simeq \int \tanh^2 \left(h + \beta \sqrt{\phi_N (\beta)} x \right) \frac{1}{\sqrt{2\pi}} e^{-x^2/2} dx.$$

From this one concludes that $\phi_N (\beta, h) \simeq q (\beta, h)$.

The fact that the m_i become approximately uncorrelated under $\mathbb{P}$, if β is small, is again proved by a contraction argument. The details of all that are quite complicated, and I refer to Talagrand's paper [25].

It is important that these kind of arguments work in much greater generality, and are applicable to many other models, for instance the perceptron, the Hopfield net, and other ones, where the interpolation method seems to be difficult to apply.

5 The Random Energy Models

5.1 The REM

This and Sect. 5.2 are a deviation from the SK-model and introduce a class of simple models invented by Derrida which in a certain vague sense are supposed to be "universal attractors" of much more complicated models like SK. On a mathematical level this is very far from being understood. Nonetheless, computations on Derrida's model (in Ruelle's asymptotic version) can be used to give a transparent proof of Guerra's bound of the free energy by the Parisi expression. We will explain this in the Sect. 5.3.

The basic difficulty of the SK-model is coming from the fact that the "energies" $H (\sigma)$ are correlated random variables. Derrida [26] realized that already something interesting is happening assuming that they are just independent random variables having (about) the correct variances. Therefore, we consider independent Gaussian random variables $(X_N (\sigma))_{\sigma \in \Sigma_N}$. Σ_N does not need to have any structure here, so we just let $1 \leq \sigma \leq 2^N$. In order to match the variance of the Hamiltonian in the SK-case, we should take $N/2$, but for convenience, we take $X_N (\sigma)$ to be i.i.d. Gaussians with variance N, put $H_{N,\beta} (\sigma) \overset{\text{def}}{=} -\beta X_N (\sigma)$, and define the partition function, the free energy, and the Gibbs measure in the usual way

$$Z_N (\beta) \overset{\text{def}}{=} \sum_\sigma e^{\beta X_N(\sigma)}, \tag{5.1}$$

$$f(\beta) \stackrel{\text{def}}{=} \lim_{N \to \infty} \frac{1}{N} \log Z_N(\beta), \quad \mathcal{G}_{N,\beta}(\sigma) \stackrel{\text{def}}{=} Z_N(\beta)^{-1} e^{\beta X_N(\sigma)}. \tag{5.2}$$

It is easy to see that the free energy is self-averaging, so that $f(\beta)$ is also the limit of the expectations, and therefore nonrandom. The Gibbs measure is again a *random* probability distribution on Σ_N, as the $H(\sigma)$ are random variables. The limiting free energy is not difficult to determine and is given by

$$f(\beta) = \begin{cases} \beta^2/2 + \log 2 & \text{for } \beta \leq \beta_{\text{cr}} = \sqrt{2 \log 2} \\ \sqrt{2 \log 2}\beta & \text{for } \beta \geq \beta_{\text{cr}} = \sqrt{2 \log 2} \end{cases}. \tag{5.3}$$

Much more interesting is the Gibbs distribution in the $N \to \infty$ limit. This can be derived from a well-known probabilistic result on extreme values of i.i.d. Gaussian random variables. There exists a sequence $a_N \uparrow \infty$ (the exact value is of no importance, they are of order $\sqrt{2 \log 2}N$) such that the random measure

$$\sum_{\sigma} \delta_{X_N(\sigma) - a_N}$$

converges weakly to a Poisson point process on $\mathbb{R}$ with intensity measure

$$\sqrt{2 \log 2} e^{-\sqrt{2 \log 2} t} dt.$$

We write $\text{PPP}(t \to a e^{-at})$ for a Poisson point process with such a density. Remark that there is a largest point, simply because $a e^{-at}$ is integrable at $+\infty$. In contrast, there is no smallest point, and the points are lying dense and denser the further one goes down the negative real axis. We can represent such a point process as $\sum_{i=0}^{\infty} \delta_{\xi_i}$, where $\xi_1 > \xi_2 > \cdots$ are real-valued random variables. We also just talk of the "point process (ξ_i)," meaning $\sum_{i=0}^{\infty} \delta_{\xi_i}$, but we tacitly always assume that the points are ordered downward. (The point processes we consider will always have a largest point.) We are not really interested in the energy levels, but rather in the Gibbs weights, which are given as $\exp[\beta X_N(\sigma)]$. As we are interested only in the relative weights, we can as well consider $\exp[\beta(X_N(\sigma) - a_N)]$. Of course, we could normalize the weights to a (random) probability distribution, but it turns out to better be not too hasty with that, and to consider first the limiting point process of these points which evidently converges in distribution to the transformation of the point process $\text{PPP}\left(t \to \sqrt{2 \log 2} e^{-\sqrt{2 \log 2} t}\right)$ obtained by applying the mapping $\xi \to \eta \stackrel{\text{def}}{=} e^{\beta \xi}$ to the points. This is a $\text{PPP}(t \to xt^{-x-1})$, with the parameter $x = x(\beta) = \sqrt{2 \log 2}/\beta$, i.e., we have

$$\sum_{\alpha} \delta_{\exp[\beta(X_N(\alpha) - a_N)]} \to \text{PPP}\left(t \to x(\beta) t^{-x(\beta)-1}\right) \tag{5.4}$$

in distribution. The $\text{PPP}(t \to xt^{-x-1})$ (which of course are point processes on the positive real line) have a number of remarkable properties which are absolutely crucial for their appearance in the Parisi picture.

Proposition 5.1. *Assume* (η_i) *are the points of a* PPP $\left(t \to xt^{-x-1}\right)$, *and let* $Y_1, Y_2, \ldots$ *be i.i.d. positive real random variables satisfying* $EY^x < \infty$, *being also independent of the point process. Set* $\psi(x) \stackrel{\text{def}}{=} (EY^x)^{1/x}$. *Then* $\sum_i \delta_{\psi(x)^{-1}Y_i\eta_i}$ *is also a* PPP $\left(t \to xt^{-x-1}\right)$. *In plain words, multiplying the points* η_i *by* Y_i *amounts to the same (when regarded as a point process) than multiplying the points with the constant* $\psi(x)$.

The proof is an easy exercise and I do not give it here. Note that the properties crucially depend on the special form of the intensity measure of the Poisson process. The property actually characterizes PPP $\left(t \to xt^{-x-1}\right)$ as has recently been shown by Ruzmaikina and Aizenman [27].

In order to describe the limiting Gibbs distribution, one still has to apply a normalization, and it is plausible that we can interchange the normalizing operation with taking the limit in (5.4), i.e., we would like to conclude that the point process $\sum_\sigma \delta_{\mathcal{G}_{N,\beta}(\sigma)}$ converges weakly to the proper normalization of PPP $\left(t \to xt^{-x-1}\right)$. There is, however, a difficulty. Let $\eta_1 > \eta_2 > \ldots > 0$ be the ordered (random) points of a PPP $\left(t \to xt^{-x-1}\right)$. We would like to apply a normalization procedure by normalizing the weights η_i, setting

$$\overline{\eta}_i = \eta_i \Big/ \sum_j \eta_j .$$

This we can only do if the sum converges. One easily proves the following statement for the points of a PPP $\left(t \to xt^{-x-1}\right)$:

$$\sum_j \eta_j < \infty \text{ a.s.} \iff x < 1.$$

If $x < 1$, we can therefore define the normalization procedure, obtaining the point process $\sum_i \delta_{\overline{\eta}_i}$ which we denote by $\mathcal{N}\left(\text{PPP}\left(t \to xt^{-x-1}\right)\right)$. This is no longer a Poisson point process as is evident from the fact that the points sum up to 1. The following result is plausible, but its proof still requires some work as the above normalization is not a continuous operation.

Proposition 5.2. *Assume* $\beta > \sqrt{2\log 2}$. *Then* $\sum_\sigma \delta_{\mathcal{G}_{N,\beta}(\sigma)}$ *converges weakly to* $\mathcal{N}\left(PPP\left(t \to xt^{-x-1}\right)\right)$, *where* $x(\beta) = \sqrt{2\log 2}/\beta$.

For a proof (in a more general setting), see [3], (Chap. 1) or [28]. The result states that for low-temperature, there are configurations σ which have Gibbs weight of order 1 in the $N \to \infty$ limit, but these Gibbs weights stay random. So the limiting Gibbs distribution is not "self-averaging." Furthermore, there is a "countable" number of such configurations in the limit. More precisely: for any $\varepsilon > 0$ there exists a number $K(\varepsilon)$ such that the total Gibbs weight of the $K(\varepsilon)$ configurations with the largest weight is $\geq 1 - \varepsilon$, with $\mathbb{P}$ probability larger than $1 - \varepsilon$, and that uniformly in N. Furthermore $K(\varepsilon)$ has to go to ∞ for $\varepsilon \to 0$. The situation is easy to understand. For $\beta > \sqrt{2\log 2}$, the Gibbs weights concentrate on the configurations σ for which the $X_N(\sigma)$ are maximal or close to the maximum. These (negative) energies (near the maximum) are spaced at a distance of order 1. The second largest is below the largest by a random distance which stays stochastically of order 1 in the $N \to \infty$ limit. The maximum is approximately at $\sqrt{2\log 2}N$, with a correction of order $\log N$.

If $\beta < \sqrt{2\log 2}$, the situation is completely different. The main contribution comes from energies approximately at a level aN, where $a < \sqrt{2\log 2}$ (actually $a = \beta$, by accident). At this level, the energies are lying tightly, with exponentially small typical spacings. Therefore, the maximum Gibbs weight for $\beta < \sqrt{2\log 2}$ is exponentially small in N, and in order to catch a macroscopic weight one has to sum over exponentially many individual configurations. Therefore, in the limit, "uncountable" many configurations contribute to the Gibbs measure.

A prediction of the Parisi theory is that the point process described above is an universal object in spin glass theory and appears as the distribution of the "pure states" in essentially all systems exhibiting "spin glass behavior," in particular in the SK-model. It is difficult to give the notion of a "pure state," which is often appearing in the physics literature, a precise mathematical sense. This has been achieved only for the p-spin SK-model which has a simpler structure than the regular SK-model, by Talagrand (see [3], Chap. 6).

5.2 The Generalized Random Energy Model

The random energy model of Sect. 5.1 is certainly on oversimplification. Derrida [29] a bit later introduced a model which has hierarchical organized correlations. Shortly afterward, Ruelle [30], in an attempt to get a clearer mathematical picture of the physicists predictions in spin glass theory, introduced a point process version, which is the limiting object of Derrida's model. This model was then further investigated in [24] and elsewhere. These models are now called "generalized random energy models," or GREM for short. In contrast to the random energy model, they have a nontrivial notion of "overlaps."

Here is Derrida's version. We consider a tree with 2^N leaves and K branching levels, where K stays fixed (for the moment), and we let then $N \to \infty$. We write the elements of the tree as $\alpha = (\alpha_1, \ldots, \alpha_K)$ where $\alpha_i \in \{1, \ldots, 2^{N/K}\}$. For convenience, we always assume that N/K is an integer. (We switch here from σs to αs as the notation for the basic configurations because σ_i could easily be mixed up with ± 1 spins.) We again write Σ_N for the collection of such αs. Evidently, we have 2^N elements in Σ_N. For $i \leq K$, we identify $(\alpha_1, \ldots, \alpha_i)$ with the "bond" from node $(\alpha_1, \ldots, \alpha_{i-1})$ to $(\alpha_1, \ldots, \alpha_i)$. To the bonds of the tree, we attach Gaussian random variables with variances proportional to N, but depending on the level inside the tree. We choose parameters $a_1^2, \ldots, a_K^2 > 0$ with $\sum_i a_i^2 = 1$, and for $i \leq K$, $(\alpha_1, \ldots, \alpha_i)$ as above, we choose Gaussian random variables $X_{\alpha_1, \ldots, \alpha_i}^{(i)}$ which have variance $a_i^2 N$. All these variables are independent. Then we define

$$X_N(\alpha) \overset{\text{def}}{=} \sum_{i=1}^{K} X_{\alpha_1,\dots,\alpha_i}^{(i)}, \tag{5.5}$$

which replaces $X_N(\sigma)$ of Sect. 5.1, i.e., for any leaf of the tree we sum the independent Gaussian variables attached to the bonds along the path from the root to this leaf.

The $X_N(\alpha)$ are evidently Gaussians with variance N, like in the REM case, but there are now correlations. Defining for $\alpha, \alpha' \in \Sigma_N$ the overlap

$$R(\alpha, \alpha') \overset{\text{def}}{=} \max\{i : (\alpha_1, \dots, \alpha_i) = (\alpha'_1, \dots, \alpha'_i)\}$$

one has

$$\mathbb{E} X_N(\alpha) X_N(\alpha') = N \sum_{i=1}^{R(\alpha, \alpha')} a_i^2.$$

We impose the following condition:

$$a_1^2 > a_2^2 > \dots > a_K^2 > 0. \tag{5.6}$$

If this is not satisfied, just some levels disappear in the $N \to \infty$ limit, so we can as well make this assumption.[1] The partition function and the Gibbs weights are defined as before in (5.1) and (5.2). Despite its simplicity the model has a number of surprising properties sketched below.

There is a sequence of critical values

$$\beta_i^{\text{cr}} \overset{\text{def}}{=} \frac{\sqrt{2 \log 2}}{\sqrt{K} a_i}.$$

If $\beta \in \left(\beta_M^{\text{cr}}, \beta_{M+1}^{\text{cr}}\right)$, $1 \le M \le K$, $\left(\beta_{K+1}^{\text{cr}} \overset{\text{def}}{=} \infty\right)$, then the Gibbs distribution "freezes" on level M but not on level $M + 1$. This means that the contribution comes from the configurations α where $X_{\alpha_1,\dots,\alpha_i}^{(i)}$ is close to its maximal value for $i \le M$, but not for $i > M$. The marginal measure of the Gibbs measure on the first M components

$$\mathcal{G}_{N,\beta}^{(M)}(\alpha_1, \dots, \alpha_M) \overset{\text{def}}{=} \sum_{\alpha_{N+1},\dots,\alpha_K} \mathcal{G}_{N,\beta}(\alpha)$$

satisfies

$$\lim_{N\to\infty} \sum_\alpha \delta_{\mathcal{G}_{N,\beta}^{(M)}(\alpha_1,\dots,\alpha_M)} = \mathcal{N}\left(\text{PPP}\left(t \to x_M t^{-x_M-1}\right)\right),$$

weakly, where

---

[1] There is a delicate issue in case of equalities in (5.6) which we do not address here.

$$x_i \stackrel{\text{def}}{=} \sqrt{2\log 2}/\sqrt{K} a_i \beta. \tag{5.7}$$

This fact is at first sight quite astonishing because it means that the limiting Gibbs distributions does not take into account the tree structure, but takes care only of the variance of the last frozen level. (For proofs of this fact, see [11, 28]).

The free energy of the GREM can be computed (see [29]), but it is not of great relevance for the aspects discussed here.

The above limit result suggests a direct tree construction of the limiting object in terms of point processes. This had been done by Ruelle [30]. The construction is at follows.

The point process PPP $\left(t \to x_1 t^{-x_1-1}\right)$ for the first level consists of countably many random points which we can assume to be ordered downward. Call them $\eta_1^1 > \eta_2^1 > \ldots > 0$. For any $i \in \mathbb{N}$, i.e., for any point from the first level, we choose independent point processes PPP $\left(t \to x_2 t^{-x_2-1}\right)$, whose points we again order downward: $\eta_{i,1}^2 > \eta_{i,2}^2 > \ldots > 0$, and in this way we proceed. For $j \le K$ and $i_1, \ldots, i_{j-1}$ fixed, $\left(\eta_{i_1 i_2 \ldots i_j}^j\right)_{i_j \in \mathbb{N}}$ are the points of a PPP $\left(t \to x_j t^{-x_j-1}\right)$. They are independent for different $i_1, \ldots, i_{j-1}$ and also for different levels j, and we again assume that for each of these point processes, the points are ordered downward. This is essentially Ruelle's cascade construction.

We can compose these point processes of the individual levels by just multiplying the "abstract Gibbs weights" along the tree (which corresponds to summing the energy levels of Derrida's GREM along the tree). We therefore arrive at a point process with random points indexed by $\mathbf{i} = (i_1, \ldots, i_K)$, $i_j \in \mathbb{N}$

$$\eta_{\mathbf{i}} = \eta_{i_1}^1 \eta_{i_1,i_2}^2 \cdot \ldots \cdot \eta_{i_1,i_2,\ldots,i_K}^K. \tag{5.8}$$

This is not a Poisson point process, but after normalization, surprisingly, it is simply a normalized REM. For convenience assume $\beta > \beta_K^{\text{cr}} \stackrel{\text{def}}{=} \sqrt{2\log 2}/\sqrt{K}\sigma_K$, so that $x_K(\beta) < 1$. (If this is not satisfied, one has to collapse some of the later levels and one arrives essentially to the same conclusion for the remaining ones.) In that case

$$\sum_{\mathbf{i}} \eta_{\mathbf{i}} < \infty$$

with probability one, and so one can normalize the point process, defining

$$\overline{\eta}_{\mathbf{i}} \stackrel{\text{def}}{=} \frac{\eta_{\mathbf{i}}}{\sum_{\mathbf{j}} \eta_{\mathbf{j}}}.$$

Then the point process

$$\Xi = \sum_{\mathbf{i}} \delta_{\overline{\eta}_{\mathbf{i}}}$$

is the normalization of a PPP $\left(t \to x_K t^{-x_K-1}\right)$.

The point process Ξ does not keep track of the way the points were produced through the tree, so it "forgets" the tree structure. This structure is, however, important for the Parisi picture. The tree structure can be retained in the following way. As usual we order the energy levels η_i downwards, i.e., we define a (random) bijection $\pi : \mathbb{N} \to \mathbb{N}^K$ such that $\eta_{\pi(k)}$ is the kth largest element in the set $\{\eta_i\}$. This leads to an overlap structure on $\mathbb{N}$, by measuring the hierarchical distance between $\pi(i)$ and $\pi(i')$, i.e., we set for $i, i' \in \mathbb{N}$

$$r(i,i') \stackrel{\text{def}}{=} \max\{r : \pi(i)_1 = \pi(i')_1, \ldots, \pi(i)_r = \pi(i')_r\}, \quad \max \emptyset \stackrel{\text{def}}{=} 0.$$

This leads to a sequence of (random) partitions of $\mathbb{N}$, which for $k \leq K - 1$ clumps together points in $\mathbb{N}$ whose π-value agrees on level k, i.e., we introduce the equivalence relation

$$i \sim_k i' \iff r(i,i') \geq k,$$

which leads to a partition $\mathcal{Z}_k$ of $\mathbb{N}$ into the equivalence classes of $\sim_k$. If k decreases, the partitions become coarser. For $k = 0$, evidently all of $\mathbb{N}$ is clumped into one set. Remarkably, this sequence of random partitions is stochastically independent of Ξ itself. Furthermore the sequence of clustering has a very simple Markovian structure, when viewed "backward." This Markovian clustering is best described in terms of a continuous time Markov process $(\Gamma_t)_{t \geq 0}$ which takes values in the compact set E of partitions of $\mathbb{N}$. If $\mathcal{Z}, \mathcal{Z}'$ are partitions of $\mathbb{N}$, we write $\mathcal{Z} \prec \mathcal{Z}'$, if $\mathcal{Z}'$ is a coarsening of $\mathcal{Z}$, i.e., if the sets in $\mathcal{Z}'$ are obtained from clumping sets in $\mathcal{Z}$. Transitions in our process are only allowed to coarser partitions, i.e., for $s < t$, we have $\Gamma_s \prec \Gamma_t$. The description of the process is most simply given in terms of traces on finite subsets I of $\mathbb{N}$. We denote the set of partitions of I by E^I. Then the trace of (Γ_t) on E^I is a Markovian process itself, with finite state space, of course, and its Q-matrix $(a^I_{\Gamma,\Gamma'})_{\Gamma,\Gamma' \in E^I}$ is given as follows: Γ' is obtained from Γ by clumping together exactly $k \geq 2$ classes into one, and if Γ has N classes, then

$$a^I_{\Gamma,\Gamma'} = \frac{1}{(N-2)\binom{N-2}{k-2}}.$$

In all other cases $a^I_{\Gamma,\Gamma'} = 0$ (when $\Gamma \neq \Gamma'$). Of course, $a^I_{\Gamma,\Gamma} = -\sum_{\Gamma' \neq \Gamma} a^I_{\Gamma,\Gamma'}$. It is easy to see that this defines the transition kernel of the process (Γ_t) uniquely. The following result is proved in [24].

Theorem 5.3. *The law of $(\mathcal{Z}_{K-1}, \mathcal{Z}_{K-2}, \ldots, \mathcal{Z}_1)$ is the same as the law of $(\Gamma_{u_1}, \ldots, \Gamma_{u_{K-1}})$, where Γ is the above Markov process starting at the partitioning into single points, and where*

$$e^{-u_i} = \frac{x_i}{x_{i+1}},$$

the x_i given by (5.7).

The above clustering process allows to compute all kind of probabilities, for instance the distribution of overlaps of two independent replicas. For instance, the $\nu^{(2)} = \mathbb{E}\mathcal{G}^{\otimes 2}$ probability that two replicas agree on level m is

$$\mathbb{E}\left(\sum_{i,j} \overline{\eta}_i \overline{\eta}_j 1_{\{r(i,j) \geq m\}}\right) = \mathbb{E}\left(\sum_{i \neq j} \overline{\eta}_i \overline{\eta}_j\right) \mathbb{P}\left(r\left(1,2\right) \geq m\right) + \mathbb{E}\left(\sum_i \overline{\eta}_i^2\right)$$

$$= \mathbb{E}\left(\sum_i \overline{\eta}_i^2\right)\left(1 - \mathbb{P}\left(r\left(1,2\right) \geq m\right)\right),$$

which may be computed explicitly, but this is of no importance here.

One crucial point is that the above reformulation does not really refer to the K levels of the GREM. One simply has a point process $(\overline{\eta}_i)$, the points ordered downward, which is a $\mathcal{N}\left(\text{PPP}\left(t \to x_K t^{-x_k-1}\right)\right)$, and independently a continuous time clustering process (Γ_t). Important therefore are only the "Gibbs weights" $\overline{\eta}_i$ and the clustering times

$$\tau_{i,j} \overset{\text{def}}{=} \inf\left\{t : i \sim_{\Gamma_t} j\right\}.$$

In this way, one steps away from the original GREM model, and in fact, the claim from the Parisi theory is that these objects appear universally in all kind of spin glasses, in particular in the SK-model. The overlaps $q\left(i,j\right)$ in the SK-model are then related to the clustering times $\tau_{i,j}$ through a decreasing function φ such that $q\left(i,j\right) = \varphi\left(\tau_{i,j}\right)$. The function φ, and the parameter x_M are the main parameters of the Parisi theory, and they are determined through the variational problem (3.5). As soon as one has determined these parameters, one can compute the distribution of the overlaps of replicas in the same way as above. The determination of the parameters is, however, not easy. For instance one does not know if the variational formula (3.5) (in the proper setup $K = \infty$ and inside the Parisi ansatz) has a unique solution.

5.3 The Aizenman–Sims–Starr Proof of Guerra's Bound of the SK Free Energy

In a remarkable paper [5], Guerra extended the bound derived in Sect. 4.1 to a bound of $f\left(\beta, h\right)$ by the Parisi solution. The proof is not very complicated, but hard to understand without knowledge of the cascade picture introduced in Sect. 5.2. A bit later, Aizenman, Sims, and Starr [31] reproved the bound, and generalized it by introducing what they call "random overlap structures," which serve as an abstract model for measures on a countable set which have a notion of "overlaps."

Definition 5.4. *A* random overlap structure *$\mathcal{R}$ (ROSt for short) consists of a finite or countable set A, a probability space $(\Gamma, \mathcal{G}, \mathbb{P})$, and random variables $\eta_\alpha \geq 0$, $q_{\alpha,\alpha'}$, $\alpha, \alpha' \in A$, satisfying the following properties:*

1. $\sum_\alpha \eta_\alpha < \infty$.
2. $(q_{\alpha,\alpha'})$ is positive definite and satisfies $q_{\alpha,\alpha} = 1$.

The η_α play the rôle of (unnormalized) Gibbs weights, and the qs are the abstract overlaps.

Example 5.5. As an example take $A = \Sigma_N \stackrel{\text{def}}{=} \{-1,1\}^N$. The η_σ, $\sigma \in \Sigma_N$, can be arbitrary. For $q_{\sigma,\sigma'}$ we take the standard overlap $R_N(\sigma,\sigma')$, as introduced before. We write $\mathcal{R}_N^{\text{SK}}$ for this overlap structure. The q here are nonrandom. On the other hand, we can use a (random) reordering of the set A by ordering the η_σ downward: $\eta_1 > \eta_2 > \ldots > \eta_{2^N}$. After this random reordering, the q become random: $q_{1,2}$ for instance is the overlap of the two indices with the largest η-weight.

Example 5.6. Another overlap structure is defined by Ruelle's probability cascades (5.8) introduced in Sect. 5.2. Fix $0 = m_0 < m_1 < \ldots < m_K = 1$. We take $A = \mathbb{N}^K$, and the η are the (unnormalized) weights η_i as in Sect. 5.2 with $x_i \stackrel{\text{def}}{=} m_i$, $1 \le i \le K$ (see (5.8)). There is a slight problem because we have to take the last parameter $x_K = 1$, which implies that $\sum_i \eta_i = \infty$. This will not cause any difficulties for what we do below. The overlaps are defined in the following way. Fix a sequence $0 \le q(1) < q(2) < \ldots < q(K) < q(K+1) = 1$, and we set

$$q_{i,i'} = q(\max\{k : (i_1,\ldots,i_k) = (i'_1,\ldots,i'_k)\} + 1),$$

i.e., we measure the hierarchical distance on the tree, and weight it with the function q. For this ROSt, we write $\mathcal{R}_K^{\text{Ruelle}}$.

Given any ROSt, we attach to it families of Gaussian random variables $(y_{\alpha,i})_{\alpha \in A, i \in \mathbb{N}}$, $(\kappa_\alpha)_{\alpha \in A}$ by requiring

$$\mathbb{E}(\kappa_\alpha \kappa_{\alpha'}) = q_{\alpha,\alpha'}^2/2, \tag{5.9}$$

and the "cavity field" by

$$\mathbb{E}(y_{\alpha,j} y_{\alpha',j'}) = q_{\alpha,\alpha'} \delta_{j,j'}. \tag{5.10}$$

The κ and the y are independent. In case, the qs themselves are random variables, these are just the conditional distributions, given (ξ,q). It is not difficult to see that such random variables exist. By an extension of the probability space, we can assume that all the random variables are defined on a single probability space.

For later use, we give the construction of the cavity variables for $\mathcal{R}_K^{\text{Ruelle}}$. We simply write

$$y_i = g^{(0)} + \sum_{k=1}^K g_{i_1,\ldots,i_k}^{(k)}, \tag{5.11}$$

where the gs are independent centered Gaussians, with var $\left(g^{(0)}\right) = q\left(1\right)$, var $\left(g^{(k)}\right) = q\left(k+1\right) - q\left(k\right)$. Furthermore, the $y_{i,j}$, $j \in \mathbb{N}$, are independent copies of y_i. The κ_i are constructed in a similar way.

The above notion of an ROSt needs some explanation. If one extends the representation (4.2) to M newcomers $\tau_1, \dots, \tau_M$, $M \ll N$, then one gets

$$-H_{N+M}\left(\sigma, \tau\right) \simeq \frac{\beta}{\sqrt{N+M}} \sum_{i<j\leq N} J_{ij}\sigma_i\sigma_j + h \sum_{i=1}^{N} \sigma_i$$
$$+ \beta \sum_{j=1}^{M} y_{\sigma,j}\tau_j + h \sum_{j=1}^{M} \tau_j,$$

where

$$y_{\sigma,j} \stackrel{\text{def}}{=} \frac{1}{\sqrt{N}} \sum_{i=1}^{N} J_{i,N+j}\sigma_i. \tag{5.12}$$

This runs exactly as in Example 5.5., but we still would like to make a replacement of $N+M$ by N in the first summand in order to compare with the Hamiltonian of the N-system. $N^{-1/2} \sum_{i<j\leq N} J_{ij}\sigma_i\sigma_j$ has variance $\left(N-1\right)/2$, and $\left(N+M\right)^{-1/2} \sum_{i<j\leq N} J_{ij}\sigma_i\sigma_j$ has variance $\left(N+M\right)^{-1} N\left(N-1\right)/2 \simeq \left(N-1\right)/2 - M/2$. We can therefore (approximately) represent the former by the latter plus an independent Gaussian $\sqrt{M/2}\kappa_\sigma$, where $\left(\kappa_\sigma\right)_{\sigma\in\Sigma_N}$ is a field with the covariances given by (5.9). If we set

$$\eta_\sigma \stackrel{\text{def}}{=} \exp\left[\frac{\beta}{\sqrt{N+M}} \sum_{i<j\leq N} J_{ij}\sigma_i\sigma_j + h \sum_{i=1}^{N} \sigma_i\right],$$

we see that

$$\frac{Z_{N+M}}{Z_N} \simeq \frac{\sum_{\sigma\in\Sigma_N,\,\tau\in\Sigma_M} \eta_\sigma \exp\left[\sum_{i=1}^{M} \left(\beta y_{\sigma,i} + h\right)\tau_i\right]}{\sum_{\sigma\in\Sigma_N} \eta_\sigma \exp\left[\beta\sqrt{M/2}\kappa_\alpha\right]}. \tag{5.13}$$

Here we have used the ROSt from the N-spin SK-model (with Gibbs weights coming from a slightly changed temperature parameter). Aizenman, Sims, and Starr had the idea to investigate the above object when the N-system is replaced by an *arbitrary* ROSt $\mathcal{R}$, and they considered the "relative finite M free energy" in the following way

$$G_M\left(\beta, h, \mathcal{R}\right) \stackrel{\text{def}}{=} \frac{1}{M}\mathbb{E}\left(\log \frac{\sum_{\alpha,\tau\in\Sigma_M} \eta_\alpha \exp\left[\sum_{j=1}^{M} \left(\beta y_{\alpha,j} + h\right)\tau_j\right]}{\sum_\alpha \eta_\alpha \exp\left[\beta\sqrt{M/2}\kappa_\alpha\right]}\right), \tag{5.14}$$

where the $\mathbb{E}$ expectation is taken with respect both to the law of the random overlap structure and the cavity variables $y_{\alpha,i}$ and κ_α. The variant of Guerra's theorem in the generalized version of Aizenman, Sims, and Starr is the following remarkable inequality, which holds for *any* M, and any ROSt.

Theorem 5.7. *For any M, and any ROSt $\mathcal{R}$ one has*

$$f_M(\beta, h) \leq G_M(\beta, h, \mathcal{R}). \tag{5.15}$$

Remark 5.8. Aizenman, Sims, and Starr actually show that

$$f(\beta, h) = \inf_{\mathcal{R}} \lim_{M \to \infty} G_M(\beta, h, \mathcal{R}),$$

which follows by plugging in the SK-ROSt itself, using (5.13), and applying the subadditivity result of Guerra–Toninelli (3.2).

I am not going to prove the theorem. To a large extent it is a rerun of the computation done above in Sect. 4.1. Here is an outline. One uses the following interpolation:

$$-H_{M,\beta}(\tau, \alpha, t) \stackrel{\text{def}}{=} \frac{\beta\sqrt{1-t}}{\sqrt{M}} \sum_{1 \leq i < j \leq M} J_{ij}\tau_i\tau_j$$

$$+ \beta\sqrt{\frac{M(1-t)}{2}}\kappa_\alpha + \beta\sqrt{t}\sum_{i=1}^{M} y_{\alpha,i}\tau_i$$

and defines

$$\hat{G}_M(\beta, h, t, \mathcal{R}) \stackrel{\text{def}}{=}$$

$$\frac{1}{M}\mathbb{E}\left(\log \frac{\sum_{\alpha \in A,\ \tau \in \Sigma_M} \eta_\alpha \exp\left[-H_{M,\beta}(\tau, \alpha, t) + h\sum_{j=1}^{M}\tau_j\right]}{\sum_{\alpha \in A} \eta_\alpha \exp\left[\beta\sqrt{M/2}\kappa_\alpha\right]}\right),$$

where $\mathbb{E}$ is taken with respect to the overlap structure *and* the Js (which are supposed to be independent). For $t = 0$, the κ-part cancels, and one just gets $f_M(\beta, h)$. For $t = 1$, one gets $G_M(\beta, h, \mathcal{R})$. By a computation similar to the one in Sect. 4.1, one gets

$$\frac{d\hat{G}_M(\beta, h,, t, \mathcal{R})}{dt} = \frac{1}{2}\nu_t^{(2)}\left((R_M(\tau, \tau') - q_{\alpha,\alpha'})^2\right) \geq 0, \tag{5.16}$$

which immediately implies the theorem. $\nu_t^{(2)}$ is to be understood in the following way. For fixed environment (from J and the ROSt), one takes two independent copies of the (τ, α) distributed according to

$$p_t(\tau, \alpha) \stackrel{\text{def}}{=} \frac{\eta_\alpha \exp\left[-H_{M,\beta}(\tau, \alpha, t) + h\sum_i \tau_i\right]}{\text{Normalization}},$$

and afterward, one takes the environment expectation.

In principle, one should of course take $\mathcal{R}_N^{\text{SK}}$ as the ROSt, in which case one gets $f(\beta, h)$ by Guerra–Toninelli in the $M \to \infty$ limit, as remarked above, but one has obviously not gained much. The striking fact, however, is that the inequality is true for *any* ROSt, and one can try to obtain good upper bounds by choosing ROSt for which one can compute G_M. It turns out that the good choice is Ruelle's $\mathcal{R}_K^{\text{Ruelle}}$ from Example 5.6..

Lemma 5.9.

$$G_M\left(\beta, h, \mathcal{R}_K^{\text{Ruelle}}\right) = G_1\left(\beta, h, \mathcal{R}_K^{\text{Ruelle}}\right) = \mathcal{P}_K\left(m, q; \beta, h\right). \qquad (5.17)$$

Proof. We give a sketch of the computation as it is not done in [31], and as it is not completely trivial. We can handle numerator and denominator in (5.14) separately. The denominator is simpler, so I will only discuss the numerator. We take $M = 1$. It will be clear from the computation that for general M the outcome is the same. With the representation of the y_i by (5.11), we get

$$\frac{1}{2}\sum_{i,\tau\in\Sigma_1}\eta_i\exp\left[(\beta y_i + h)\tau\right] = \sum_i\eta_i\cosh\left(\beta y_i + h\right)$$

$$= \sum_{(i_1,\ldots,i_K)}\eta_{i_1}^1\eta_{i_1i_2}^2\cdots\cdots\eta_{i_1i_2\ldots i_K}^K\cosh$$

$$\left(\beta\sum_{n=0}^K g_{i_1,\ldots,i_n}^{(n)} + h\right). \qquad (5.18)$$

We condition on $\eta_{i_1}^1, \eta_{i_1i_2}^2, \ldots, \eta_{i_1i_2\ldots i_{K-1}}^{K-1}$ and $g_{i_1}^{(1)}, g_{i_1i_2}^{(2)}, \ldots, g_{i_1i_2\ldots i_{K-1}}^{(K-1)}$. Then $\left(\eta_{i_1i_2\ldots i_K}^K\right)_{i_K\in\mathbb{N}}$ is a PPP $\left(t \to m_K t^{-m_K-1}\right)$ whose points are multiplied by the independent random variables $\left(\cosh\left(\beta\sum_{n=0}^K g_{i_1,\ldots,i_n}^{(n)} + h\right)\right)_{i_K}$. From Proposition 5.1., we know that the conditional law (conditioned on anything up to level $K - 1$) of the point process

$$\left(\eta_{i_1i_2\ldots i_K}^K\cosh\left(\beta\sum_{n=0}^K g_{i_1,\ldots,i_n}^{(n)} + h\right)\right)_{i_K}$$

is the same as that of

$$\left(C_K\left(\beta\sum_{n=0}^{K-1} g_{i_1,\ldots,i_n}^{(n)}\right)\eta_{i_1i_2\ldots i_K}^K\right)_{i_K},$$

where

$$C_K\left(\xi\right) \stackrel{\text{def}}{=} \left[E_Z\cosh^{m_K}\left(\xi + h + \beta\sqrt{q_{K+1} - q_K}Z\right)\right]^{1/m_K}, \quad \xi \in \mathbb{R},$$

Z being a standard Gaussian random variable, and E_Z the expectation with respect to Z. C_K is a random variable which still depends on the $g^{(n)}$, $n \leq K - 1$. We proceed in the same way, replacing $\left(C_K\left(\beta\sum_{n=0}^{K-1} g_{i_1,\ldots,i_n}^{(n)}\right)\eta_{i_1i_2\ldots i_{K-1}}^{K-1}\right)_{i_{K-1}\in\mathbb{N}}$ by

$$\left(C_{K-1}\left(\beta\sum_{n=0}^{K-2} g_{i_1,\ldots,i_n}^{(n)}\right)\eta_{i_1i_2\ldots i_{K-1}}^{K-1}\right)_{i_{K-1}\in\mathbb{N}},$$

where

$$C_{K-1}(\xi) \overset{\text{def}}{=} \left[E_Z C_K^{m_{K-1}} \left(\xi + h + \beta \sqrt{q_{K-1} - q_{K-2}} Z \right) \right]^{1/m_{K-1}},$$

and so on. We finally see that multiplying the points η_i by $\cosh(\beta y_i + h)$ amounts (for the corresponding point process) in multiplying the points by the constant $E \log Y_1$ from (3.12). The denominator in (5.14) is simpler because there one has in every step just an integration of a Gaussian in the exponent. We therefore see that multiplying the points η_i by $\exp\left[(\beta/\sqrt{2}) \kappa_i \right]$ simply leads to a multiplication of the point process by $\exp\left[(\beta^2/4) \sum_{i=0}^{K} m_i \left(q_{i+1}^2 - q_i^2 \right) \right]$. In the definition of G_1 (5.14), we would now like to argue that $\sum_i \eta_i$ cancels out. There is the slight difficulty that this sum diverges almost surely, because of $m_K = 1$, but we can choose m_K slightly less than 1, in which case the sum is finite, and so cancels, and then we can let $m_K \to 1$ in the end. The upshot of this computation is that

$$G_1(\beta, h, \mathcal{R}) = E \log Y_1 - \frac{\beta^2}{4} \sum_{i=0}^{K} m_i \left(q_{i+1}^2 - q_i^2 \right) + \log 2$$
$$= \mathcal{P}_K(m, q; \beta, h),$$

the $\log 2$ is coming from dividing by 2 in (5.18). It is evident from this computation that we get the same for arbitrary M. (One is just having M factors of $\cosh(\cdot)$ with independent contents, so in every step of the above argument, the factoring remains.) $\square$

Combining this result with Theorem 5.7., one gets

$$f_M(\beta, h) \leq \mathcal{P}_K(m, q; \beta, h)$$

for any K, and any sequence m and q. Therefore

$$f_M(\beta, h) \leq \inf_{K, m, q} \mathcal{P}_K(m, q; \beta, h).$$

This is Guerra's upper bound.

6 Some Open Problems: Ultrametricity and Chaos

One of the puzzling open problems is the ultrametricity. In the physics literature, people often speak of "pure states" which are organized in a hierarchical way. The concept of a pure state is difficult to make precise for mean-field models. A way to make the concept precise is by the so-called "metastates" introduced by Newman and Stein. I do not want to discuss this concept here in details, and refer to the literature. For the SK-model, there are no results in this direction. If I understand the physics literature correctly, it is roughly claimed that the configurations for a large N-system can be lumped into groups such that inside the groups two independent choices (under the Gibbs measure) have the same overlaps (with probability $\simeq 1$). These groups are then called the "pure states." Of course, these groups depend on the random couplings. The Gibbs weights of the groups are random itself, and are given by the Poisson–Dirichlet point process as in the GREM. The groups (i.e., the pure states) can then be lumped into supergroups, in such a way that independent choices of configurations in different groups but the same supergroup have a constant overlap. Then these supergroups can be clumped into supersupergroups, etc. On a mathematical level, this has been made precise in the case of the p-spin SK by Talagrand, but in this case, there are only the groups, and there is only one supergroup which consists of all the configurations. One can find the analysis in Talagrand's book ([3], Chap. 6). For the SK-model itself, there should be a continuous hierarchy, which is difficult to make precise, but see the contribution in this volume by Bovier and Kurkova [11] for a similar construction in the GREM case. For the SK-model, there is no mathematically rigorous description for these concepts available.

On a more modest level, the first task would be to prove that the configurations satisfy, with probability close to 1, an ultrametricity property. If we choose three independent configurations $\sigma, \sigma'\sigma''$ under the Gibbs distribution, then all $\varepsilon > 0$

$$\lim_{N \to \infty} \mathbb{E}\left(\mathcal{G}^{\otimes 3}\left\{R_N\left(\sigma, \sigma''\right) \geq \min\left(R_N\left(\sigma, \sigma'\right), R_N\left(\sigma', \sigma''\right)\right) - \varepsilon\right\}\right) = 1,$$

where $\mathcal{G}^{\otimes 3}$ the threefold product measure of the Gibbs measure, applied to $(\sigma, \sigma'\sigma'')$. There is no proof of this, yet, which is the more astonishing as the Parisi solution in physics literature is derived with a hierarchical structure in mind.

The failure to prove this means that some of the core problems in (mean-field) spin glasses are still very poorly understood, mathematically. (Of course, it may be that ultrametricity is wrong, but this would be even more puzzling.) There probably is a basic principle which tells that in a large class of models, anything which is not ultrametrically organized is suppressed in the $N \to \infty$ limit.

In order to get some (modest) insight into this property, we have recently proved such a statement in a nonhierarchical version of the GREM [32, 33]. Here is this version.

Consider the set $I = \{1, \ldots, n\}$, as well as a collection of positive real numbers $\{a_J\}_{J \subset I}$ such that

$$\sum_J a_J = 1.$$

For convenience, we put $a_\emptyset = 0$. The relevant subset of J will be only the ones with positive a-value. For $A \subset I$ we set

$$\mathcal{P}_A \stackrel{\text{def}}{=} \{J \subset A : a_J > 0\}, \quad \mathcal{P} \stackrel{\text{def}}{=} \mathcal{P}_I.$$

For $N \in \mathbb{N}$, we set $\Sigma_N \overset{\text{def}}{=} \{1, \ldots, 2^N\}$. We also fix positive real numbers γ_i, $i \in I$, satisfying

$$\sum_i \gamma_i = 1,$$

and write $\Sigma_N^i \overset{\text{def}}{=} \Sigma_{\gamma_i N}$, where for notational convenience we assume that $\gamma_i N$ is an integer. We will label the "spin configurations" α as

$$\alpha = (\alpha_1, \ldots, \alpha_n), \ \alpha_i \in \Sigma_N^i,$$

i.e., we identify Σ_N with $\Sigma_N^1 \times \cdots \times \Sigma_N^n$. For $J \subset I$, $J = \{j_1, \ldots, j_k\}$, $j_1 < j_2 < \cdots < j_k$, we write $\Sigma_{N,J} \overset{\text{def}}{=} \prod_{s=1}^k \Sigma_N^{j_s}$, and for $\alpha \in \Sigma_N$, we write α_J for the projected configuration $(\alpha_j)_{j \in J} \in \Sigma_{N,J}$. Our Hamiltonian is defined as

$$-H_{N,\beta}(\alpha) \overset{\text{def}}{=} \beta \sum_{J \in \mathcal{P}} X_{\alpha_J}^J, \tag{6.1}$$

where $X_{\alpha_J}^J$, $J \in \mathcal{P}$, $\alpha_J \in \Sigma_{N,J}$, are independent centered Gaussian random variables with variance $a_J N$. The X_α are then Gaussian random variables with variance N, but they are correlated. $\mathbb{E}$ will denote expectation with respect to these random variables. The GREM corresponds to the special situation where the sets in $\mathcal{P}$ are "nested," meaning that $\mathcal{P}$ consists of an increasing sequence of subsets. Without loss of generality we may assume that in this case

$$\mathcal{P} = \{J_m : 1 \leq m \leq k\}, \ J_m \overset{\text{def}}{=} \{1, \ldots, n_m\}, \tag{6.2}$$

where $1 \leq n_1 < n_2 < \cdots < n_k \leq n$. In the GREM case, the natural metric coming from the covariance structure

$$d(\alpha, \alpha') \overset{\text{def}}{=} \sqrt{\mathbb{E}\left((X_\alpha - X_{\alpha'})^2\right)}$$

is an ultrametric. In the more general case (6.1) considered here, this metric is evidently not an ultrametric.

Any of our models can be "coarse-grained" in many ways into a GREM. For that consider strictly increasing sequences of subsets of I: $\emptyset = A_0 \subset A_1 \subset \cdots A_K = I$. We do not assume that the A_i are in $\mathcal{P}$. We call such a sequence a chain $\mathbf{T}$. We attach weights $\hat{a}_{A_j}$ to these sets by putting

$$\hat{a}_{A_j} \overset{\text{def}}{=} \sum_{B \in \mathcal{P}_{A_j} \setminus \mathcal{P}_{A_{j-1}}} a_B.$$

Evidently $\sum_{j=1}^K \hat{a}_{A_j} = 1$, and if we assign random variables $H_{\mathbf{T},N}(\alpha)$, according to (6.1), we arrive after an irrelevant renumbering of I at a GREM of the form (6.2). In particular, the corresponding metric d is an ultrametric. The partition function, free energy, and the Gibbs measure are defined in the usual way

$$Z_N(\beta) = \sum_\alpha \exp\left[-H_{N,\beta}(\alpha)\right], \ F_N(\beta) = \frac{1}{N}\log Z_N(\beta),$$

$$f_N(\beta) = \mathbb{E}\left(F_N(\beta)\right),$$

$$\mathcal{G}_{N,\beta}(\alpha) = \frac{1}{Z_N(\beta)}\exp\left[-H_{N,\beta}(\alpha)\right].$$

For any chain **T**, we attach to our model a GREM-Hamiltonian $(H_{\mathbf{T},N}(\alpha))_{\alpha\in\Sigma_N}$, as explained above, and arrive at the corresponding free energy.

Theorem 6.1.
$$f(\beta) \stackrel{\text{def}}{=} \lim_{N\to\infty} f_N(\beta)$$

exists, and is also to almost sure limit of $F_N(\beta)$.

$f(\beta)$ is the free energy of a GREM. More precisely, there exists a chain **T** *such that*
$$f(\beta) = f(\mathbf{T},\beta), \ \beta \geq 0.$$

$f(\mathbf{T},\beta)$ is minimal in the sense that

$$f(\beta) = \min_{\mathbf{S}} f(\mathbf{S},\beta),$$

the minimum being taken over all chains.

For a proof, see [32]. Under some mild nondegeneracy hypothesis,[2] it is also possible to really prove the ultrametricity in this model. In our model, the natural notion of an overlap $R(\alpha,\tau)$ of two configurations is the subset of I where they agree. Therefore, the overlap R takes values in the set of subsets of I. The ultrametricity then states that

$$\lim_{N\to\infty} \mathbb{E}\left(\mathcal{G}^{\otimes 3}\left(R(\alpha,\alpha')\cap R(\alpha',\alpha'') \subset R(\alpha,\alpha'')\right)\right) = 1.$$

For the GREM, one has of course $R(\alpha,\alpha')\cap R(\alpha',\alpha'') \subset R(\alpha,\alpha'')$ for all $\alpha,\alpha'\alpha''$ by construction, but for the above model, this is not the case. So the above statement is a nontrivial result. For the exact conditions, and the statement, see [33].

A problem which is equally puzzling as the ultrametricity is the so-called "chaos property." As that is explained in Talagrand's contribution to this volume [12], I am not discussing it here.

---

[2] It is somewhat complicated to state precisely, so I do not give it here. Essentially it rules out such degenerate cases as where the Hamiltonian is the sum of independent Hamiltonians depending on the individual σ_i.

7 Other Models: Some Comments about the Perceptron

The Parisi theory applies to many other problems besides to the SK-model, e.g., to the assignment problem from combinatorial optimization, to the perceptron and the Hopfield net from neural networks, to coding theory, and to others. Many of these applications are presented in Nishimori's book [21]. I will give some comments about the perceptron which I find particularly interesting. The perceptron is a problem coming up in the theory of neural nets. In the language of combinatorics, it is the problem of how many points of $\{-1,1\}^N \subset \mathbb{R}^N$ do belong to the intersections of M random half-planes. We can describe the random half-planes by

$$U_k \stackrel{\text{def}}{=} \left\{ x \in \mathbb{R}^N : \left\langle x, g^{(k)} \right\rangle \geq 0 \right\},$$

where $\langle \cdot, \cdot \rangle$ denotes the standard inner product, and $g^{(k)}$ are standard Gaussian vectors in $\mathbb{R}^N$. We set

$$\mathcal{N}(N, M) \stackrel{\text{def}}{=} \# \left(\{-1,1\}^N \cap \bigcap_{k=1}^{M} U_k \right).$$

The main interest is in the case $M = \alpha N$, $\alpha > 0$, and $N \to \infty$. As $\mathbb{E}\mathcal{N}(N, M) = 2^{N-M}$, only the case $\alpha < 1$ is interesting. It is believed that there is a critical value $\alpha^* < 1$ such that $\mathcal{N}(N, \alpha N) \to 0$ for $\alpha > \alpha^*$, and

$$f(\alpha) = \lim_{n \to \infty} \frac{1}{N} \log \mathcal{N}(N, \alpha N) > 0$$

for $\alpha < \alpha^*$. Talagrand has been able to prove a formula for $f(\alpha)$ for small α, and to prove that $\mathcal{N}(N, \alpha N) \to 0$ for α close to 1. This can be found in his book [3] (Chaps. 3 and 4).

I am not going to explain any of these results here, but only give an indication why the problem is closely connected to SK-type models, and actually also to the cavity approach. We have

$$\mathcal{N}(N, M) = \lim_{\beta \to \infty} \sum_{\sigma \in \{-1,1\}^N} \exp \left[-\beta \sum_{k \leq M} 1_{\{\sigma \notin U_k\}} \right].$$

From this representation, it looks being natural to investigate this problem first for finite β, and $N, M \to \infty$, hoping of course that one can interchange the limit with $\beta \to \infty$ in the end. One may then even replace the indication function with a smoother function $u : \mathbb{R} \to \mathbb{R}$, and investigate (incorporating β into u) Hamiltonians of the form

$$-H_{N,M}(\sigma) \stackrel{\text{def}}{=} \sum_{k=1}^{M} u \left(\frac{1}{\sqrt{N}} \sum_{i=1}^{N} g_{i,k} \sigma_i \right), \tag{7.1}$$

where the $g_{i,k}$ are standard Gaussian variables. The quantities inside u are exactly the cavity variables $y_{i,\sigma}$ introduced in (5.12), but here we still apply a nonlinear function u to them. A Hamiltonian of the above form appears in a somewhat "simplified" SK-model (which, however, is not really much simpler than the original one), where there are two group of spin variables $\sigma_1, \ldots, \sigma_N, \tau_1, \ldots, \tau_N$, and the σs interact with the τs but where there is no internal interaction inside the two groups. This would mean that one has a Hamiltonian

$$-H_N(\sigma, \tau) \stackrel{\text{def}}{=} \frac{\beta}{\sqrt{N}} \sum_{i,j=1}^{N} J_{ij} \sigma_i \tau_j.$$

When calculating the partition function, one can of course sum out one of the groups, take for instance the τs

$$Z_N = \sum_{\sigma} 2^N \prod_{j=1}^{N} \cosh\left(\frac{\beta}{\sqrt{N}} \sum_{i=1}^{N} J_{ij} \sigma_i\right) = 2^N \sum_{\sigma} \prod_{j=1}^{N} \cosh(\beta y_{j,\sigma})$$

$$= 2^N \sum_{\sigma} \exp\left[\sum_{j=1}^{N} \log \cosh(\beta y_{j,\sigma})\right],$$

with the cavity variables $y_{j,\sigma}$ defined by (5.12). Therefore, one is exactly back to (7.1). As remarked above, I am not going into any details of the analysis of Talagrand here, but I want to make just one final comment why I think that an understanding of these models might be at the core getting a better understanding of mean-field spin glass theory.

We can write (7.1) in a more complicated way, by introducing the "empirical" cavity field

$$L_{M,\sigma} \stackrel{\text{def}}{=} \frac{1}{M} \sum_{j=1}^{M} \delta_{y_{j,\sigma}}$$

which are random elements in the space of probability measures on $\mathbb{R}$. Then

$$-H_{N,M}(\sigma) = M \int u(x) L_{M,\sigma}(dx),$$

but one may define more general nonlinear Hamiltonians. Given a real-valued function Φ defined on the space of probability measures, we set

$$-H_{N,M,\Phi} \stackrel{\text{def}}{=} M\Phi(L_{M,\sigma}),$$

and one may ask about the asymptotic behavior of the free energy

$$f_{N,M,\Phi} \stackrel{\text{def}}{=} \frac{1}{N} \log \sum_{\sigma} \exp[M\Phi(L_{M,\sigma})].$$

The perceptron is the special case where Φ is linear. As far as I can see, a theory for the general case is lacking completely. What Talagrand has been able to do is to give (partial) answers for linear Φ. For "classical" mean-field type problems (much more general than the Curie–Weiss problem), the investigation of empirical measures and their large deviation behavior had been conceptually very important. One might hope that something similar is correct for mean-field type spin glasses. Here is a (somewhat imprecisely formulated) question, I would find interesting to answer. Take $M = N$ for simplicity. For a fixed σ, $L_{N,\sigma}$ is the empirical measure of N i.i.d. standard Gaussian variables, and therefore one knows from standard large deviation theory that for a probability measure μ on $\mathbb{R}$ which is different from the standard normal distribution ϕ, one has

$$\mathbb{P}\left(L_{N,\sigma} \sim \mu\right) \sim \exp\left[-NI\left(\mu\right)\right]$$

with the "good" rate function

$$I\left(\mu\right) = \int \log \frac{d\mu}{d\phi} d\mu.$$

(I am using here the standard large deviation jargon.) Therefore, the expected number of σs with $L_{N,\sigma} \sim \mu$ is $2^N \exp\left[-NI\left(\mu\right)\right]$, which is exponentially growing, provided μ is close enough to ϕ, but that will not be the typical number of σs. The question: is it true that for μ sufficiently close to ϕ, one is in a "replica-symmetric" situation, which means for instance that the typical σs which satisfy $L_{N,\sigma} \sim \mu$ have components which are approximately independent (conditionally on the environment coming from the cavity variables), with conditional means m_i, which are itself approximately i.i.d. under the law of the environment. Even more challenging would of course be to have a full Parisi theory for arbitrary μ, not just the ones close to ϕ, where one would expect a spin glass behavior.

8 Dynamic Problems and Aging

Gibbs distributions of the type (2.1) are usually thought to be the equilibrium measures for a Markovian time evolution. In continuous time, the evolution $\{p_t\}$ is described by the so-called Q-matrix $q\left(\sigma,\tau\right) \overset{\text{def}}{=} dp_t\left(\sigma,\tau\right)/dt|_{t=0}$. Typically, one is looking for a reversible dynamics, i.e., one that satisfies the detailed balance equations

$$\mathcal{G}\left(\sigma\right)q\left(\sigma,\tau\right) = \mathcal{G}\left(\tau\right)q\left(\tau,\sigma\right).$$

A particular case is given by

$$q\left(\sigma,\tau\right) = \left[1 + \exp\left[-H\left(\sigma\right) + H\left(\tau\right)\right]\right]^{-1}, \ \sigma \neq \tau.$$

Under mild conditions, one knows that for a finite system, one has convergence to the equilibrium measure which is a "justification" to consider the equilibrium measure only, because after sufficiently long time, this is the distribution to consider anyway. However, for large systems, the time necessary to reach this (approximately) may be very long, and it is therefore of crucial importance to know how long it really takes till one can safely neglect that one has started outside equilibrium. There is a huge literature about this topic, and the problem is particularly challenging for disordered systems. It is in fact generally agreed that in spin glasses, the necessary time is too long, and the nonequilibrium situation is the more relevant for practical purposes.

A quantity of crucial importance is the time–time correlation, for instance given by

$$C_N\left(t, s\right) \overset{\text{def}}{=} \left|\Sigma_N\right|^{-1} \sum_\sigma R_N\left(\sigma\left(t\right), \sigma\left(t + s\right)\right),$$

where $\sigma\left(t\right)$ is the spin configuration at time t for the process starting at $\sigma\left(0\right) = \sigma$, and its limit

$$C\left(t, s\right) \overset{\text{def}}{=} \lim_{N \to \infty} C_N\left(t, s\right).$$

For a system, developing in a "classical" way, one expects that for large t, this quantity becomes independent of s. This means, that the system decouples in a constant way, regardless how old it is, i.e., how large t is. (Of course, there is in any case a limit $t \to \infty$ involved as one cannot expect this independence of s after a very short time.) However, this is not the way spin glasses do behave. They typically develop slower the older they are. Intuitively, this means that the system become trapped in metastable parts of the configuration space from which it takes more and more time to escape. This kind of analysis has become a topic of intensive research, both in theoretical physics and in mathematics, and it goes without saying that in the theoretical physics literature there are many predictions, based on nonrigorous theoretical investigations and numerical approaches, which cannot be proved yet in mathematical terms. A prediction which goes under the name "aging" is a behavior like

$$C\left(t, s\right) \simeq \text{const} \times \left(s/t\right)^\alpha.$$

Such a behavior has been mathematically analyzed and proved in a number of cases, but not for the SK-model. Here are a number of models for which it has been shown to occur.

- *Dynamics of the random energy model.* This has been a topic of intensive research, mainly by Ben Arous, Bovier, and Gayrard [34]. One of the aspects which makes the problem very challenging is that fact that one regards a dynamic which mimics in some aspects the natural dynamics of the SK-model, namely one identifies $1, \ldots, 2^N$ with $\{-1, 1\}^N$, and allows jumps only to nearest neighbor on the hypercubic lattice. This introduces a structure for the dynamics which is absent in the original REM-model, and the advantage is that results have a better chance to give insights into what happens in more realistic models.

- *Spherical SK-models.* The simplification here is coming from replacing the spin state $\{-1,1\}$ by a continuous sphere. The mathematical advantage is that one can then diagonalize the matrix (J_{ij}), and use powerful tools from random matrix theory. Investigations on this model have begun with work of Ben Arous, Dembo, and Guionnet [35], and is explained here in the survey article by Guionnet [13].
- *Random trap models.* These had been introduced by Bouchaud [36]. There are several versions of it. One may for instance look at an ordinary random walk in $\mathbb{Z}$ or $\mathbb{Z}^d$ but with the modification that the sites are traps which are randomly placed (i.i.d. for instance). Also the depths of the traps are random, and the deeper the trap is, the longer has the random walk to remain in the trap. If the tail of the trap distribution is sufficiently thick, then the model exhibits aging. This has been investigated by Ben Arous and Cerny [37].

References

1. Sherrington, D. and Kirkpatrick, S.: *Solvable model of a spin glass.* Phys. Rev. Lett. **35** (1972), 1792–1796
2. Mézard, M., Parisi, G. and Virasoro, M.A.: Spin *glass theory and beyond.* World Scientific, Singapore, 1987
3. Talagrand, M.: Spin *glasses: a challenge for mathematicians.* Springer, Berlin Heidelberg, New York, 2003
4. Guerra, F. and Toninelli, F.L.: *The thermodynamic limit in mean field spin glass models.* Commun. Math. Phys. **230** (2002), 71–79
5. Guerra, F.: *Replica broken bounds in the mean field spin glass model.* Commun. Math. Phys. **233** (2003), 1–12
6. Talagrand, M.: *The Parisi formula.* Ann. Math. **163** (2006), 221–263
7. Parisi, G.: *A sequence of approximate solutions to the S-K model for spin glasses.* J. Phys. A **13**, (1980), L–115
8. Sherrington, D.: *Spin glasses: a perspective.* In this volume
9. Newman, Ch. and Stein, D.L.: *Local versus global variables in spin glasses.* In this volume
10. Newman, Ch. and Stein, D.L.: *Short range spin glasses: results and speculations.* In this volume
11. Bovier, A. and Kurkova, I.: *Much ado about Derrida's GREM.* In this volume
12. Talagrand, M.: *Mean field models of spin glasses: some obnoxious problems.* In this volume
13. Guionnet, A.: *Dynamics for spherical models of spin glass and aging.* In this volume
14. Georgii, H.O.: *Gibbs measures and phase transitions.* de Gruyter 1988
15. Aizenman, M. and Wehr, J.: *Rounding effects of quenched randomness on first-order phase transitions.* Commun. Math. Phys. **130** (1990), 48–9528
16. Bricmont, J. and Kupiainen, A.: *Phase transition in the 3d random field Ising model.* Commun. Math. Phys. **116** (1988) 539–572
17. Ledoux, M.: *The concentration of measure phenomenon.* AMS 2001

18. Aizenman, M., Lebowitz, J. and Ruelle, D.: *Some rigorous results on the Sherrington–Kirkpatrick model.* Commun. Math. Phys. **112** (1987), 3–20
19. Fröhlich, J. and Zegarlinski, B.: *Some comments on the Sherrington–Kirkpatrick model of spin glasses.* Commun. Math. Phys. **112** (1987), 553–566
20. de Almeida, J.R.L. and Thouless, D.J.: *Stability of the Sherrington–Kirkpatrick solution of spin glasses.* J. Phys. A **11** (1978), 983
21. Nishimori, H.: *Statistical physics of spin glasses and information processing.* Oxford Science Publications, Oxford, 1999
22. Comets, F.: *A spherical bound for the Sherrington–Kirkpatrick model.* Asterisque **236** (1998), 103–108
23. Guerra, F. and Toninelli, F.L.: *Quadratic replica coupling in the Sherrington–Kirkpatrick mean field spin glass model.* J. Math. Phys. **43** (2002), 3704–3716
24. Bolthausen, E. and Sznitman, A.-S.: *On Ruelle's probability cascades and an abstract cavity method.* Commun. Math. Phys. **197** (1998), 247–276
25. Talagrand, M.: *The Sherrington–Kirkpatrick model: a challenge to mathematicians.* Probab. Theory Rel. Fields **110** (1998), 109–176
26. Derrida, B.: *Random energy model: an exactly solvable model of disordered systems.* Phys. Rev. B **24** (1981), 2613–2626
27. Ruzmaikina, A. and Aizenman, M.: *Characterization of invariant measures at the leading edge for competing particle systems.* Preprint (2005)
28. Bovier, A. and Kurkova, I.: *Derrida's Generalized Random Energy Models I & II.* Annales de l'Institut Henri Poincaré **40** (2004), 439–495
29. Derrida, B.: *A generalization of the random energy model that includes correlations between the energies.* J. Phys. Lett. **46** (1986), 401–407
30. Ruelle, D.: *A mathematical reformulation of Derrida's REM and GREM.* Commun. Math. Phys. **108** (1987), 225–239
31. Aizenman, M., Sims, R. and Starr, S.L.: *An extended variational principle for the SK spin-glass model.* Phys. Rev. B **68** (2003), 214403
32. Bolthausen, E. and Kistler, N.: *On a non-hierarchical version of the Generalized Random Energy Model.* Ann. Appl. Probab. **16** (2006), 1–14
33. Bolthausen, E. and Kistler, N.: *On a non-hierarchical version of the Generalized Random Energy Model. Part II. Ultrametricity.* Preprint (2006)
34. Ben Arous, G., Bovier, A. and Gayrard, V.: *Glauber dynamics of the random energy model. 1. Metastable motion on the extreme states.* Commun. Math. Phys. **235** (2003), 379–425
35. Ben Arous, G., Dembo, A., and Guionnet, A.: *Aging in spin glasses.* Probab. Theory Rel. Fields **120** (2001), 1–67
36. Bouchaud, J.-P.: *Weak ergodicity breaking and aging in disordered systems.* J. Phys. I (France) **2** (1992), 1705–1713
37. Ben Arous, G. and Cerny, J.: *Bouchaud's model exhibits two different aging regimes in dimension one.* Ann. Appl. Probab. **15** (2005), 1161–1192

Spin Glasses: A Perspective

David Sherrington

Rudolf Peierls Centre for Theoretical Physics, University of Oxford
1 Keble Rd., Oxford OX1 3NP, United Kingdom
e-mail: d.sherrington1@physics.ox.ac.uk

Summary. A brief personal perspective is given of issues, questions, formulations, methods, some answers and selected extensions posed by the spin glass problem, showing how considerations of an apparently insignificant and practically unimportant group of metallic alloys stimulated an explosion of new insights and opportunities in the general area of complex many-body systems and still is doing so.

1 Introduction

What are spin glasses? The answer to this apparently innocuous question has evolved from an initially obscure, if interesting, small special class of metallic alloys to one concerned with the globally pervasive issue of the understanding of emergent complex behaviour in many-body systems, the development of new mathematical, simulational and conceptual tools, new experimental protocols, new algorithms and even a new class of mathematical probability problems. In this paper I shall review some of this history and try to expose some of the key issues, challenges, solutions and opportunities of the topic.

2 Random Magnetic Alloys

The story starts with magnetic aspects of metallic alloys. In the early 1960s there was much interest in the solid-state physics community in the behaviour of isolated impurities in metals, first with the formation of local magnetic moments on magnetic metal impurities in non-magnetic hosts [1] and then with the strong coupling of a localized moment to the conduction electrons at low temperatures and its consequences for the electrical resistivity [2]. The later 1960s and early 1970s saw the emergence of interest in the effects of inter-impurity correlations through spin glasses (see e.g. [3,4]) and the Kondo lattice [5].

The appellation "spin glass" is due to Bryan Coles in the late 1960s to label the low temperature state of a class of substitutional magnetic alloys, typified by **Cu**Mn or **Au**Fe, with finite concentrations of the magnetic ions Mn or Fe in the non-magnetic hosts Cu and Au. The reason for the name is twofold, first that in the state the magnetic moments (traditionally called "spins") on the magnetic ions seem to freeze in orientation but without any periodic ordering (so conceptually reminiscent of the amorphous freezing of the locations of atoms in a conventional (structural) glass), and secondly that the low temperature specific heat is linear in T, again a feature of conventional glasses. Experiments at that time indicated a non-sharp onset of the state as the temperature was reduced from the paramagnetic one, suggesting a rapid onset of sluggishness but not a phase transition, again as believed to be characteristic of glasses. There were attempts to understand the behaviour in the 1960s but nothing very extensive.

But then more accurate experiments in the early 1970s exposed a new source of theoretical interest, an apparently sharp phase transition signalled by a cusp in the magnetic susceptibility when external magnetic fields were kept very small [6]. This had to be a new type of phase transition and therefore worthy of extra notice. But still theoretical work was minimal until Edwards and Anderson [7] produced a paper that at one fell swoop recognized the importance of the combination of frustration and quenched disorder as fundamental ingredients, introduced a more convenient model, a new and novel method of analysis, new types of order parameters, a new mean field theory, new approximation techniques and the prediction of a new type of phase transition apparently explaining the observed susceptibility cusp. This paper was a watershed.

Edwards and Anderson's new approach was beautifully minimal, fascinating and attractive but also their analysis was highly novel and sophisticated, involving radically new concepts and methods but also unusual and unproven ansätze, as well as several different approaches. And so it seemed sensible to look for an exactly soluble model for which their techniques could be verified. Such a model was suggested by Sherrington and Kirkpatrick [8]. It extends the Edwards–Anderson model, in which exchange interactions are range-dependent and effectively short-range, to one with interactions between all spins, chosen randomly and independently from an intensive distribution (and so "infinite-ranged" but not uniform). It offered the possibility of exact solution in the thermodynamic limit and an exact mean field theory, in analogy but subtle extension of the infinite-range ferromagnet for which naïve mean field theory is correct. Study of the SK model, or the mean-field theory of the EA model that it defines, has proven highly non-trivial and instructive, and opened many new conceptual doors. It has also proven to be an entry point to many applications and extensions, which are still ongoing.

2.1 More Details

Experimental Spin Glasses

Let me be more explicit. The original experimental spin glasses can be characterized by Hamiltonians

$$H = -\frac{1}{2} \sum_{i,j} J(\mathbf{R}_i - \mathbf{R}_j) \mathbf{S}_i \cdot \mathbf{S}_j, \tag{2.1}$$

where the i, j label magnetic ions with Heisenberg spins $\mathbf{S}_i$ and locations $\mathbf{R}_i$ and $J(\mathbf{R})$ is an exchange interaction which oscillates in sign as a function of the spin separation. In metallic systems the origin of $J(\mathbf{R})$ is the Ruderman–Kittel–Kasuya–Yoshida (RKKY) interaction. In the original alloys the disorder is substitutional on a lattice.

Edwards–Anderson

What Edwards and Anderson (correctly) surmised was that the important aspect of (2.1) is the combination of frustration, corresponding to the fact that the spins receive conflicting relative ordering instructions (as a consequence of the oscillation of the exchange with separation), and the quenched disorder in the location of the spins. For theoretical convenience they effectively replaced the Hamiltonian by one that can be written as

$$H = -\frac{1}{2} \sum_{i,j} J_{ij} \mathbf{S}_i \cdot \mathbf{S}_j, \tag{2.2}$$

with spins on all the sites of a lattice but the J_{ij} between neighbouring spins and chosen randomly from a distribution having weight of either sign.[1] They further chose the distribution to be Gaussian of zero mean, thereby both eliminating the possibility of any conventional order (with spatially uniform or periodic magnetization) and also having a minimal one-parameter characterization.[2] This necessitated the introduction of a new form of order parameter to describe magnetic freezing without periodicity. In fact EA gave two versions: one based explicitly on temporal freezing

$$q = \lim_{t \to \infty, \tau \to \infty} q(t, t+\tau); \quad q(t, t+\tau) \equiv N^{-1} \sum_i \langle \mathbf{S}_i(t) \cdot \mathbf{S}_i(t+\tau) \rangle, \tag{2.3}$$

---

[1] In fact the EA Hamiltonian was first written explicitly by Sherrington and Southern [9].

[2] Actually, the Gaussian choice for the single-parameter description was also useful for the further analytic methods employed. The alternative simple single-parameter symmetric distribution having two delta functions of equal weight at $\pm J$ has often been employed in (later) computer simulations.

where $\langle \cdot \rangle$ refers to a dynamical average, and the other based on ensemble-averaging,

$$q = N^{-1} \sum_i |\langle \mathbf{S}_i \rangle|^2, \tag{2.4}$$

with $\langle \cdot \rangle$ now referring to an ensemble-average restricted to one symmetry-breaking macro-state. The phase transition is signalled by q becoming non-zero.

EA did not attempt a full solution but used several new variants of mean field theory, all requiring novel treatment beyond those conventional for a simple ferromagnet. The most sophisticated of them introduced and employed the so-called "replica trick" which replaces the average of the logarithm of the partition function,[3] the physical generating function,[4] by the limiting behaviour of a partition function of an effective periodic system of higher dimensional spins

$$\overline{\ln Z} = \lim_{n \to 0} \partial/\partial n (\overline{Z^n}) = \lim_{n \to 0} \partial/\partial n (\overline{\prod_{\alpha=1,\dots,n} Z(\alpha)}) = \lim_{n \to 0} \partial Z_{\mathrm{eff}}(n)/\partial n, \tag{2.5}$$

where the overbar refers to the average over the distribution of the J, Z is the usual partition function $Z = \mathrm{Tr}_{\{\sigma\}} \exp\{-\beta H\}$, $Z(\alpha)$ is the partition function for spins with dummy labels α and $Z_{\mathrm{eff}}(n)$ is the partition function of a periodic Hamiltonian $H_{\mathrm{eff}}(\{\sigma_i^\alpha\})$ of effectively higher dimensional pseudo-spins with extra replica labels $\alpha = 1, \dots, n$ and higher order interactions now between spins with different replica, as well as site, labels. Within this new description EA devised a new mean field theory with a new order parameter measuring inter-replica overlap

$$q^{\alpha\beta} = N^{-1} \sum_i \mathbf{S}_i^\alpha \cdot \mathbf{S}_i^\beta; \quad \alpha \neq \beta. \tag{2.6}$$

To go further, however, they employed several assumptions and approximations whose validity was difficult to assess, although they do yield results with some qualitative similarity to several experimental features.

Sherrington–Kirkpatrick

In view of the many uncertainties of the EA analysis and the fact that the model was surely not soluble with current techniques, it seemed sensible to look for a model in which a mean-field theory might be exact. Since the

---

[3] The argument for studying the average of the logarithm of the partition function is that the physical quantities it generates should be self-averaging, independent in the thermodynamic limit of the specific instance of choice of the disorder.

[4] Certain physical observables can already be expressed as derivatives of $\ln Z$ with Z defined as above with the bare H. Others, in principle, require the addition to H of terms involving appropriate conjugate fields so that desired observables follow from derivatives of $\ln Z$ with respect to these fields.

conventional ferromagnet is soluble in the thermodynamic limit provided that all spins interact equally with one another and correspondingly the exchange interaction scales inversely with the number of spins, it seemed reasonable to look for an analogue in the spin glass problem. This led to the formulation of the Sherrington–Kirkpatrick (SK) model whose Hamiltonian is similar to that of EA but with interactions between all spins, chosen randomly and independently from a distribution whose mean and variance scale inversely with the number of spins. Simplifying to Ising spins and allowing for a ferromagnetic bias and an external field, the SK model is characterized by[5]

$$H = - \sum_{(ij)} J_{ij}\sigma_i\sigma_j - h\sum_i \sigma_i; \quad \sigma = \pm 1; \quad J_{ij} \text{ i.i.d.}; \quad \overline{J_{ij}} = J_0/N, \overline{J_{ij}^2} = J^2/N.$$

(2.7)

Despite its apparent simplicity, this model has turned out to expose many subtleties; for statistical physics, for mathematical physics and for probability theory; as well as having much wider application relevance. Extensions to other related models with extensive and super-extensive constraints independently drawn from identical (intensive) distributions have led to further novelties and applications. In this introductory perspective I shall restrict discussion to outlines at the level of conceptual theoretical statistical physics, leaving mathematical rigour to other authors.

Replica Theory

Within the replica theory of EA but applied to the SK model the averaged free energy can be expressed in a form

$$\overline{F} = -T\overline{\ln Z} = -T\lim_{n\to 0}\frac{\partial}{\partial n}\mathop{\mathrm{Tr}}_{\{\sigma_i^\alpha;\alpha=1,\dots,n\}}\exp\left\{f\left(\sum_{i\alpha}\sigma_i^\alpha, \sum_{i(\alpha\beta)}\sigma_i^\alpha\sigma_i^\beta\right)\right\}, \quad (2.8)$$

in which f involves the spin variables only in the form of sums over all sites and those sums only up to quadratic order. Hence, by introducing auxiliary (macroscopic) variables to linearize these sums, the trace over the spins may be taken to yield

$$\overline{F} = -T\overline{\ln Z} = -T\lim_{n\to 0}\frac{\partial}{\partial n}\int \prod_{\alpha=1,\dots,n} dm^\alpha \prod_{(\alpha\beta)} dq^{\alpha\beta} \exp[-N\Phi(\{m^\alpha; q^{\alpha\beta}\})]$$

(2.9)

with Φ intensive. Thus, provided the limit $n \to 0$ and the thermodynamic limit $N \to \infty$ can be inverted, the method of steepest descents in principle yields a solution determined by an extremum of Φ. However, to take the limit $n \to 0$ an appropriate analytic form continuable to non-integer n is needed

---

[5] We use notation (ij) to denote a pair of unequal sites.

and the correct way to achieve this is not obvious. EA and SK both used the natural "replica-symmetric" ansatze,

$$m^\alpha = m, \text{ all } \alpha; \quad q^{\alpha\beta} = q, \text{ all } \alpha \neq \beta.^6 \tag{2.10}$$

This ansatz also yields the identifications

$$m = \overline{\langle\sigma_i\rangle} \text{ and } q = \overline{|\langle\sigma_i\rangle|^2}. \tag{2.11}$$

Already it gives many features qualitatively similar to ones found experimentally. In fact, though, it does not in general give a stable solution [10] and a much more subtle replica-symmetry-breaking ansatz for q is needed to yield stability in all regions of control-parameter space. The Parisi ansatz [11] has satisfied this need and passed all subsequent stability tests.

Let me first describe Parisi's ansatz in terms of its original replica theory formulation and only turn later to its physical interpretation. $q^{\alpha\beta}$ can be viewed as an $n \times n$ matrix with zeros on its diagonal elements.[7] The Parisi ansatz may be viewed as the result of a sequence of operations in which (1) $n(\equiv m_0)$ is initially considered as an integer which is subdivided sequentially into an integral number of smaller intervals; first into n/m_1 blocks of size m_1, then each of the m_1 blocks into m_1/m_2 blocks of size m_2 and so on sequentially, with all the m_i integers and the ratios m_i/m_{i+1} also integers, until $m_{k+1} = 1$ (2) $q^{\alpha\beta}$ is taken to have the value q_i if $I(\alpha/m_i) = I(\beta/m_i)$, $I(\alpha/m_{i+1}) \neq I(\beta/m_{i+1})$ where $I(x)$ is equal to the smallest integer greater than or equal to x, as illustrated below for the sequence $n = m_0 = 12$; $m_1 = 4$; $m_2 = 2$; $m_3 = 1$

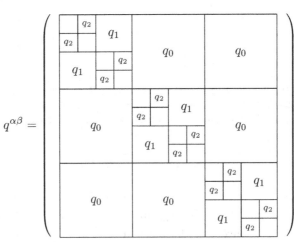

[6] One also requires that the extremum of Φ be a minimum for the single-replica order parameter m^α but a maximum for the two-replica order parameter $q^{\alpha\beta}$.

[7] Some authors take $q^{\alpha\alpha}$ as unity (cf. an extension of (2.6) but here I am) assuming the $\alpha\alpha$ term is so taken explicitly.

(3) in the limit $n \to 0$ the m are continued to real values with $0 \leq m_1 \leq m_2 \leq \cdots \leq m_k \leq m_{k+1} \equiv 1$ and q is replaced by a function $q(x)$ given by $q(x) = q_i$; $m_i < x < m_{i+1}$ $(i = 1, \ldots, k)$ with x in $[0,1]$ and (4) the limit $k \to \infty$ is taken. Insertion into (2.9) yields a functional integral which in the limit $N \to \infty$ is extremally dominated and yields self-consistency equations for the dominating $q(x)$; hereafter $q(x)$ is taken to refer to this extremal function, which is the the mean-field order function for the problem. For different regions of the (J, J_0, h, T) parameter space the stable solutions are of one of two forms:

1. $q(x) = q = $ constant; replica-symmetric (RS)
2. $q(x) = q_0$ for $0 \leq x \leq x_1$, monotonically increasing smoothly between x_1 and x_2, and $q(x) = q_1$ for $x_2 \leq x \leq 1$; full replica symmetry breaking (FRSB).[8]

They are separated by a plane in (J, J_0, h, T) which marks a continuous transition, with RSB on the higher-J side.

Replica symmetry breaking signals the existence of many non-equivalent macrostates. $q(x)$ provides a measure of the extent of similarity between these states. It follows from consideration of the concept of overlaps [12]. The overlap between two macrostates S, S' is defined by $q^{SS'} = N^{-1} \sum_i \langle \sigma_i \rangle^S \langle \sigma_i \rangle^{S'}$, where $\langle \cdot \rangle^S$ refers to a thermodynamic average over macrostate S, and the distribution of overlaps is given by $P(q) = \sum_{S,S'} W_S W_{S'} \delta(q - q^{SS'})$, where W_S is the probabilistic weight of state S, given in equilibrium by $W_S = \exp(-\beta F_S)/\sum_{S'} \exp(-\beta F_{S'})$, where F_S is the free energy of macrostate S. The relation to $q(x)$ is $\overline{P(q)} = \int dx \delta(q - q(x)) = dx/dq$. Consequently it follows that an RS system has a single macrostate (aside from trivial global inversion or rotation), whereas FRSB implies a hierarchy of non-equivalent relevant macrostates at the temperature of interest.[9] This in turn implies that the macroscopic dynamics will be slow and glassy and that practical equilibration will be very difficult to achieve. Already, however, the existence of RSB predicts different kinds of response functions; for the susceptibility one may experience either single-macrostate response $\chi_{SS} \equiv \chi_{EA} = T^{-1}(1 - q(1))$ or the full Gibbs average $\chi_G = T^{-1}(1 - \int dx q(x))$. These in turn can be identified with the experimental zero-field-cooled and field-cooled susceptibilities and used to explain their difference in the spin glass phase (see e.g. [13]); this non-ergodicity was already observed before EA in the difference between thermoremanent and isothermal remanent magnetizations (e.g. [14]). The Parisi replica analysis also demonstrates a number of other interesting properties

---

[8] For the intermediate approximations mentioned above one would have a k-step replica symmetry-breaking with $k + 1$ sections of constant $q(x)$ separated by k discontinuities, but it is believed that the only stable situations for the SK model are $k = 0$ (RS) and $k = \infty$ (FRSB). There are stable 1 step RSB solutions for several other problems (see Sect. 5).

[9] There are even more macrostates of relevance at different temperatures.

[15], such as ultrametricity [16][10] and non-self-averaging of certain non-trivial overlap measures. but these will not be dwelt upon further here.[11]

Short-Range Spin Glasses

"Real" experimental spin glasses have short-range or spatially decaying exchange interactions, whereas the replica theory above is exact only for infinite-range problems. Many of the predictions of mean field theory are mimicked qualitatively in the experiments; some are thought to be real, but others are still subjects of controversy in true Gibbs equilibrium although often apparent as non-equilibrium experimental features. The Edwards–Anderson model with nearest neighbour interactions is considered representative of such real spin glasses but remains without full exact solution.

Spin Glasses on Dilute Random Networks

A class of model spin glasses with finite inter-spin connectivities, as is the case for EA, but range-free and offering the possibility of exact solution, was introduced by Viana and Bray [18] and characterized by an analogue of SK with

$$H = - \sum_{(ij)} c_i c_j J_{ij} \sigma_i \sigma_j; \text{ random quenched } c \text{ and } J_{ij}; c_i = 0, 1; \overline{J_{ij}}$$

$$= J_0, \ \overline{J_{ij}^2} = J^2, \tag{2.12}$$

where the annealed spins σ are located on the quenched vertices of a finite-connectivity Erdős–Renyi[12] graph, with $c_i = 1$ denoting a vertex, but without the need for inverse N-scaling of the exchange distribution.[13] This problem, which is often considered a "half-way house" between SK and EA, requires more order parameters m^α, $q^{\alpha\beta}$, $q^{\alpha\beta\gamma}$, $q^{\alpha\beta\gamma\delta}$,... and, although soluble in RS approximation via a mapping $q^{\alpha\beta...r} = \int P(h)\{\tanh(\beta h)\}^r$, also poses greater challenges than SK for FRSB (see e.g. [20]).

---

[10] This corresponds to the hierarchical clustering of overlaps as illustrated by the branching cartoon

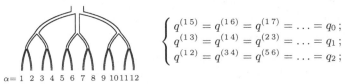

$$\begin{cases} q^{(1\,5)} = q^{(1\,6)} = q^{(1\,7)} = \ldots = q_0 \, ; \\ q^{(1\,3)} = q^{(1\,4)} = q^{(2\,3)} = \ldots = q_1 \, ; \\ q^{(1\,2)} = q^{(3\,4)} = q^{(5\,6)} = \ldots = q_2 \, ; \end{cases}$$

$\alpha = 1\ 2\ 3\ 4\ 5\ 6\ 7\ 8\ 9\ 10\,11\,12$

[11] For a recent review of the topic of overlaps and their interpretation see [17].

[12] In an Erdős–Renyi graph of degree p any vertex is connected to any other with a probability p/N.

[13] A simple extension utilizes as underlying network a random graph with fixed degree at each vertex [19].

Itinerant Spin Glasses

Thus far we have discussed only systems with magnetic moments even in the absence of interaction. However, it is well known that ferromagnetism in periodic systems can occur not only through the orientation of effectively pre-existing localized moments, as typified by Curie–Weiss mean field theory and found in insulating magnets and in some rare earth metals, but also through the spontaneous cooperative ordering of metallic itinerant electrons, as in Stoner–Wohlfarth ferromagnetism in transition metals. Similarly, one can readily envisage itinerant spin glass behaviour [21] and indeed it is found in alloys such as **RhCo** [22]. A simple model is given by the Hamiltonian

$$H = \sum_{ij\sigma} t_{ij} a_{i\sigma}^+ a_{j\sigma} + \sum_i V_i a_i^+ a_i + \sum_i U_i n_{i\uparrow} n_{i\downarrow}, \qquad (2.13)$$

where the a_i, a_i^+ are Wannier electron creation and annihilation operators, $n_{i\sigma} \equiv a_{i\sigma}^+ a_{i\sigma}$ are number operators, and the parameters t_{ij}, V_i, U_i depend upon the types of atom at sites i, j. The simplest instance takes randomly quenched alloys with two atomic types (A, B) but with the t_{ij} independent of the atom types and considers only magnetic fluctuations

$$H = \sum_{ij\sigma} t_{ij} a_{i\sigma}^+ a_{j\sigma} - \frac{1}{2} \sum_i U_i (n_{i\uparrow} - n_{i\downarrow})^2, \qquad (2.14)$$

in which the U_i take one value U_A at the sites associated with atom type A and take the another value U_B at the sites associated with atom type B. Of particular interest is the itinerant case in which (1) A is not spontaneously magnetically ordered, i.e. $(1 - U_A \chi^0(q)) > 0$, where $\chi^0(q)$ is the wave-vector dependent susceptibility associated with the bare band structure, (2) the pure B system is spontaneously itinerantly ferromagnetic (so $(1 - U_B \chi^0(0)) < 0$), but also and (3) there is no magnetic moment associated with an isolated B atom in an A matrix, even in the mean field sense of Anderson [1]. Analogy with the phenomenon of Anderson localization [23] leads to the expectation of statistical fluctuation nucleation of cluster moments within the conceptual framework of Anderson local moment formation, while further cluster interaction can stabilize cluster glass behaviour beyond a critical B concentration [21], as well as ferromagnetism at a higher concentration. However, in fact isolated paramagnetic cluster moments are not necessary precursors for spontaneous spin glass order (as emphasised by Hertz [24]), as neither there are local moments in pure itinerant ferromagnets nor well-defined bosons in BCS superconductivity above their respective onset temperatures.

A classical mean field theory follows from (2.14) by formulating the partition function as a functional integral over a Grassmann representation of the electron field [51], linearizing the interaction term involving the U over auxiliary Hubbard–Stratonovich fields, integrating out the electron fields and taking the static approximation. This yields an effective classical field theory with

$$Z = \int D\mathbf{m} \exp(-\beta \mathbf{F}(\mathbf{m})), \qquad (2.15)$$

where

$$F(\mathbf{m}) = \sum_i U_i m_i^2 - \sum_{ij} U_i U_j m_i m_j \chi_{ij}^0 - \sum_{ijkl} U_i U_j U_k U_l m_i m_j m_k m_k \Lambda_{ijkl}^0 + ..., \qquad (2.16)$$

where $m_i = < n_{i\uparrow} - n_{i\downarrow} >$ and χ_{ij}^0 , Λ_{ijkl}^0, etc. are two-, four-, etc. -point correlation functions of the bare band structure (in real space). Taking the extremum yields a set of self-consistent mean field equations which are the analogue of the Thouless–Anderson–Palmer [25] (TAP) equations for the SK equation. The analogy with Anderson localization follows from writing these equations as

$$U_i^{-1} M_i - \sum_j \chi_{ij}^0 M_j - \sum_{jkl} \Lambda_{ijkl}^0 M_j M_k M_l + .. = 0; M_i = U_i m_i \qquad (2.17)$$

and comparing with the Anderson wave-function localization equation

$$\epsilon_i \phi_i + \sum_j t_{ij} \phi_j - E \phi_i = 0, \qquad (2.18)$$

with disorder in the $\epsilon_i (= U_i^{-1})$; naively, local moment clusters of (2.17) are related to negative energy states of (2.18) and long-range magnetic order is related to the mobility edge. But in fact there are more subtle effects, both bootstrap effects as mentioned earlier (contained in the non-linear terms of (2.17)) and effects differentiating spin glass and ferromagnetic cooperative order.

A simple conceptual model of itinerant spin glass ordering, further simplified in the EA spirit, is given by an effective field theory with

$$Z = \int \prod_i d\phi_i \exp(-F(\phi_i)); F(\phi) = r \sum_i \phi_i^2 + u \sum_i \phi_i^4 - \sum_{(ij)} J_{ij} \phi_i \phi_j; u > 0, \qquad (2.19)$$

with the J_{ij} random as in EA or SK; this model encompasses local moment spin glasses for $r < 0$ and itinerant spin glasses for $r > 0$.

Other Induced Moment Models

There are other classical models allowing the bootstrapping of magnetic order. One such is the spin glass analogue of the induction of ferromagnetism due to exchange interaction lifting of singlet ground state preference of isolated atoms. A simple example is the spin-1 Ising model

$$H = -D \sum_i S_i^2 - \frac{1}{2} \sum_{ij} J_{ij} S_i S_j; \quad S_i = 0, \pm 1. \tag{2.20}$$

If $D > 0$ then the system behaves analogously to the usual spin 1/2 Ising model, but if $D < 0$ then in the absence of J the ground state preference is for non-magnetic $S_i = 0$. However, even if $D < 0$ a sufficient exchange can self-consistently lift the preference to the magnetically ordered state via a first-order transition. If the J_{ij} are quenched random as in the SK model, this system is known as the Ghatak–Sherrington (GS) model [26] and has induced spin glass behaviour; it has been analysed extensively in FRSB by Crisanti and Leuzzi [27]. The Fermionic Ising Spin Glass model [28] is closely related [29].

Vector Spin Glasses

Magnetic alloy spin glasses are not restricted to Ising systems. Indeed Heisenberg magnets are more common experimentally. It is straightforward to extend the exactly soluble models to encompass vector spins (see e.g. [30]). In the absence of a magnetic field or a ferromagnetic component there is little change of note beyond the extension of random spin glass freezing to the full spin dimensionality.[14] Within the infinite-range/ mean-field model, an axial symmetry-breaking due to an applied field or to ferromagnetism still permits a spin glass freezing in the orthogonal directions [31] with strong onset of transverse non-ergodicity and induced weaker longitudinal non-ergodicity, crossing over to strong RSB in all directions at a lower temperature [32–34]. Anisotropy effects can also be included [35].

3 Discontinuous Transitions

For the case of conventional two-spin interactions, as employed in both the SK and EA models and believed to be appropriate for conventional experimental magnetic alloy spin glasses, mean field theory yields full replica symmetry breaking once the spin glass state occurs.[15] However, in extensions which lack reflection and definiteness symmetries, such as p-spin models for $p > 2$ [36] or Potts or quadrupolar spin glasses beyond critical Potts or vector dimensions

---

[14] But it might be noted that the choice of Ising spins by SK led to the "smoking gun" of negative entropy at T=0 and the realization that there was a subtlety, which eventually led to Parisi's ansatz and beyond; negative entropy at T=0 was a known pathology of continuous classical spins but should not occur for discrete spins such as Ising.

[15] Although it remains controversial as to whether any RSB holds in short-range systems.

[37, 38] one finds that the spin glass transition is discontinuous to one step of replica symmetry breaking with finite overlap magnitude (D1RSB).[16] This behaviour is thought to be characteristic also of (even short-range interaction) structural glasses, in which crystallization is dynamically avoided in favour of self-consistent glassiness.

4 Beyond Magnetic Alloys

4.1 Complex Many-Body Problems

The formalism and concepts developed for model magnetic alloys have found significant application more generally; in particular for a large class of problems that can be characterized by control functions of the form

$$H = H(\{J_{ij...k}\}, \{S_{ij...l}\}, \{X\}), \tag{4.1}$$

where the i, j are microscopic identification labels; the $\{J_{ij...k}\}$ symbolize a set of quenched parameters depending on one or more of the identification labels and in general different for different labels; the $\{S_{ij...l}\}$ symbolize the (annealed) microscopic variables again depending on one or more identification variables; and the $\{X\}$ are macroscopic intensive control variables. The specific identifications of the $\{J, S, X\}$ can, however, be quite different, as also the manner of operation of the control function. In the spirit of statistical physics and probability theory one often concerns oneself with problems in which the $\{J, X\}$ are drawn from intensive distributions independent of the specific labels.

Examples

We have already seen one example in the case of a magnet with the i labelling the spins, the J exchange interactions, the S spin orientations, X the temperature and H the Hamiltonian determining the distribution of the S through the Boltzmann measure. Other examples include:

1. *The Hopfield neural network.* Here the i label neurons, $\{S_i\}$ indicate the states of the neurons as firing or not firing, $\{J_{ij}\}$ label synaptic efficacies given in terms of (randomly chosen quenched) stored patterns $\{\xi_i^\mu\}$; $\mu = 1, \ldots, p = \alpha N$ by $J_{ij} = N^{-1} \sum_\mu \xi_i^\mu \xi_j^\mu$, $X \equiv T \equiv \beta^{-1}$ is a measure of the rounding of the sigmoidal response of a neuron to the sum

---

[16] In a Potts or quadrupolar model for a range of intermediate Potts or vector dimensions the transition to 1RSB is continuous [37–40]; a similar transition to C1RSB occurs in a $p >$ two-spin model in a sufficient applied field [36]. Except for spherical spins, there is also a lower temperature transition from 1RSB to FRSB [41, 42].

of its incoming signals, $H = -\frac{1}{2}\sum_{ij} J_{ij} S_i S_j$ and $P(\{S\}) \sim \exp(-\beta H)$ characterizes the stationary macro-firing states. From the neural retrieval perspective, however, interest is not in the full Gibbs average but rather in the individual retrieval macrostates with macroscopic overlaps $m^\mu = N^{-1}\sum_i \xi_i^\mu \langle S_i \rangle$ with the patterns coded in the $\{J\}$; retrieval corresponds to a finite overlap with a single pattern and is an analogue of ferromagnetism in the examples of Sect. 2 Spin glass states do occur due to pattern interference but are not the desired states in neural operation and their dominance indicates breakdown of retrievable memory.[17]

2. *Hard optimization.* Here the objective is to minimize a cost function H as a function of variables $\{S\}$ with constraints $\{J\}$. An example is the problem of partitioning the vertices i of a random graph into two groups of equal size but with the minimum number of edges of the graph between the two groups. This can be formulated as finding the ground state of a Viana–Bray-like spin glass. Consequently it can be studied by an analogue of the procedure of studying the thermodynamics of the VB spin glass. If the interest is in finding the average minimum spanning cut then replica procedure may be employed, inventing an artificial annealing temperature T and taking it to zero at the end of the calculation. Of course the actual calculation involves all the subtleties of replica symmetry breaking and computer simulation involves all the corresponding issues of slow glassy dynamics.[18] Another optimization problem in artificial neural network theory is to determine the maximum number of patterns which can be stored and retrieved with a specified maximum error; in this case the variables are the synaptic efficacies and the quenched parameters are the stored patterns. More recently many other optimization problems have been studied by techniques derived from spin glass studies.

3. *Error-correcting codes.* One procedure for coding and retrieval is to code the information to be transmitted in the form of exchange interactions whose insertion into an effective magnetic Hamiltonian yields a ground state which identifies the desired message. In practice, however, transmission lines add noise and retrieval is required to best eliminate the effects of the noise. This yields yet another optimization problem, with best retrieval resulting from the introduction of an effective retrieval temperature–noise matching that on the line.[19] Indeed there are several other problems in which the optimal character of noise matching can be demonstrated.

---

[17] For further discussion see for example [43] or [44].

[18] Simulation also exhibits the spin glass features of ultrametricity and non-self-averaging [19].

[19] Again see [44] for further details.

5 Dynamics

Thus far discussion has been about equilibrium or quasi-equilibrium. However, often one wishes to consider dynamics, including away from equilibrium;[20] indeed if detailed balance is not present one cannot use usual Boltzmann equilibrium theory. As before, we are normally interested in systems characterized by simple distributions. Again one can utilize the general picture of a controlling function as in (4.1) but now operating in an appropriate microscopic dynamics (and without necessarily symmetries such as $J_{ij} = J_{ji}$). The analogue of the use of the partition function for thermodynamics is to use a dynamical generating functional [45] which can be expressed symbolically either, for random sequential updates, as

$$Z(\Lambda) = \int D\mathbf{S}(t)\Pi\delta \text{ (eqn of motion) } \exp\left(\int dt\Lambda(t) \cdot \mathbf{S}(t)\right), \qquad (5.1)$$

where the integral is over all variable paths in the full space–time, the $\Pi\delta$ term indicates that the microscopic equations of motion are always satisfied and the $\Lambda \cdot \mathbf{S}$ term is a generating term, or, for parallel updates, as

$$Z(\Lambda) = \int \Pi d\mathbf{S}(t) \prod_t W(\mathbf{S}(t+1)|\mathbf{S}(t))P_0(\mathbf{S}) \exp\left(\sum_t \Lambda(t) \cdot \mathbf{S}(t)\right), \qquad (5.2)$$

where $P_t(\mathbf{S})$ indicates the ensemble distribution of $\mathbf{S}$ at time t and $W(\mathbf{S}(t+1)|\mathbf{S}(t))$ indicates the probability of updating from $\mathbf{S}(t)$ to $\mathbf{S}(t+1)$. With suitable Jacobian normalization (not shown explicitly) $Z(\Lambda = 0) = 1$ and one can average over the quenched disorder without need for replicas; instead of interactions between replicas one gets effective interactions between different epochs. In the case of range-free problems one can again eliminate microscopic variables in place of macroscopic ones by the artifice of introducing new variables via relations such as

$$1 = \int dC(t,t')\delta(C(t,t') - N^{-1}\sum_i S_i(t)S_i(t'))$$

$$= \int d\hat{C}(t,t')dC(t,t') \exp\left\{i\hat{C}(t,t')(C(t,t') - N^{-1}\sum_i S_i(t)S_i(t'))\right\} \qquad (5.3)$$

and similarly for response functions (involving also operators corresponding to $\partial/\partial S_i(t)$). One can then integrate out the microscopic variables to leave purely macroscopic measures; in the simplest cases of the form

$$Z_{\text{eff}} \sim \int D\widetilde{C}(t,t',t'',\dots) \exp(N\Phi(\{\widetilde{C}\})), \qquad (5.4)$$

---

[20] Note that whereas in real physical situations the true microscopic dynamics is determined by nature, in computer simulations the dynamics is chosen by the simulator and there exists the opportunity to optimise that choice. Similarly, the control fields X are choosable.

where $\widetilde{C}$ is used to denote the generic set and the temporal dependence is two-time for full connectivity of the SK type but includes all numbers of different times for VB finite connectivity. Steepest descents then yields self-consistent coupled equations for the macroscopic correlation and response functions, although of course boundary conditions need care. This is the dynamical analogue of replica thermodynamics. In general, however, it is more difficult than replica theory and fewer cases have been solved fully. Also, in some cases the convenience of a final expression purely in terms of coupled correlation and response functions is not available, although alternative descriptions in terms of ensembles of effective single agents can often be obtained.

An alternative procedure invoking an infinite multiplicity of single-time order parameters has also been considered but will not be pursued here (see e.g. [46])

5.1 Examples

p-Spin Spherical Spin Glass

One example of the above procedure has been the analysis of the (infinite-range) $p(> 2)$-spin spherical spin glass, of Hamiltonian

$$H = - \sum_{i_1 < i_2 < \cdots < i_p} J_{i_1 i_2 \ldots i_p} S_{i_1} S_{i_2} \ldots S_{i_p}; \quad \sum_i S_i^2 = N; \quad \overline{J_{i_1 i_2 \ldots i_p}^2} = J^2 p! / 2 N^{p-1}$$

(5.5)

and obeying Langevin dynamics, for which closed equations in terms of correlation and response functions have been obtained [47]. In general these equations are not restricted to stationarity. Analysis has indicated that above a critical temperature, known as the dynamical transition temperature, stationary solutions do exist, with $\widetilde{C}(t, t') \equiv \widetilde{C}(t - t')$ and satisfying the normal fluctuation–dissipation theorem and mode-coupling theory, but below this temperature equilibration does not occur, the normal fluctuation–dissipation theorem $-dR/dC = \beta$ (where R is the integrated response, C is the correlation function and β is the inverse temperature) is replaced by a modified relation $-dR/dC = \beta X(C)$ where $X(C) = x(C)$ with $x(q)$ the inverse of the Parisi function $q(x)$, the R and C are now two-time (and non-stationary) and the C-dependence of $X(C)$ is instantaneous parenthetic.[21, 22]

These and related dynamical studies vindicate and quantify the concepts of glassy dynamics deduced from the thermodynamic existence of many non-equivalent metastable macrostates and the barriers between them.

---

[21] See for example [17].

[22] In this case the onset of RSB is discontunuous and the transition temperature is given differently by simple extremization of replica theory and dynamically. The correct comparison with dynamics within replica theory is determined using marginal stability.

Dynamical SK-Model

In the $p(> 2)$-spin spherical spin glass model there is only one step of RSB in the replica equilibrium theory and similarly only two straight slope regions for $X(C)$. The $p = 2$ SK Ising system is more complicated with more structure, corresponding to the hierarchy of FRSB, and dynamical analogues of ultrametricity [48]. Other models can show regions of 1-RSB and of FRSB thermodynamics[23] while it seems likely that dynamical vestiges of FRSB may occur in many systems, even with 1RSB thermodynamics.

Minority Game

A rather different example is found in the so-called Minority Game in econophysics (see e.g. [49, 50]), which mimics a system of speculative agents in a model market trying to gain by minority action. In the batch version of this game the system obeys microdynamics

$$p_i(t+1) = p_i(t) - h_i - \sum_j J_{ij} \operatorname{sgn} p_j(t); \; h_i = N^{-1} \sum_j \xi_i \cdot \omega_j; \; J_{ij} = N^{-1}\xi_i \cdot \xi_j,$$

(5.6)

with the i labelling agents, the p unbounded variables corresponding to strategy point-weightings, and the $\vec{\xi}$ and $\vec{\omega}$ quenched random vectors in a D-dimensional strategy space. This system is soluble along the lines outlined above, utilizing large-N steepest descent domination, in terms of an ensemble of independent agents obeying non-Markovian stochastic dynamics with ensemble-self-consistently determined coloured noise. On the macroscopic level it exhibits an ergodic–nonergodic transition at a critical value d_c of $d = D/N$, asymptotically independent of preparation for $d > d_c$ but preparation-dependent for $d < d_c$.

6 Conclusion

The spin glass problem has yielded many new concepts and techniques in both theoretical and experimental physics. These concepts and techniques have in turn inspired new insights and practical opportunities in the wider field of complex many-body problems, ranging through physics, computer science, biology and economics, with pastures still open in these and the social sciences. Most of this work has been on simple models with a single level of microscopic timescale (but many resulting macro timescales) but some work has started and much remains to do when different parameters are allowed different microdynamic time-scales (as for example in neural networks where both neurons and synapses evolve, the former much faster than the latter,

---

[23] For example the $p > 2$-spin Ising model has 1-RSB thermodynamic behaviour above a critical temperature but then FRSB behaviour below; see e.g. [41,42].

or in biological evolution where the timescales of organism operation and species evolution and mutation are very different). Although physical systems normally obey detailed balance, others need not (e.g. biological or economic or social systems). Most of the theoretical work has been performed at a level of uncertain if physically reasonable approximation or assumption. Greater mathematical physics rigour is now needed and will be the topic of other authors in this volume. The spin glass models have introduced also new concepts in probability theory that are stimulating new mathematics. Spin glass dynamics poses challenges yet to be investigated with mathematical rigour. Much has been achieved but much remains to do.

7 Acknowledgements

My career in spin glasses has involved and benefited from many valuable collaborations and discussions with too many people to list, but they know who they are and I thank them all. I have also benefited from financial support from UK Engineering and Physical Sciences Research Council (and its predecessors), the European Science Foundation, the European Community and the Royal Society, as well as from Imperial College, the University of Oxford and other universities and institutions.

I would like to thank explicitly Dr Isaac Pérez Castillo, for a careful reading and correction of drafts of this paper and Dr Andrea Sportiello for providing the drawings.

References

1. Anderson P.W., Phys. Rev. **124**, 41 (1961)
2. Kondo J., J. Prog. Theor. Phys. **32**, 37 (1964)
3. Coles B.R., The origins and influences of the spin glass problem in "Multicritical Phenomena" eds. R. Pynn and A. Skjeltorp (Plenum, New York, 1983)
4. Mézard M., Parisi G. and Virasoro M.A., "Spin Glass Theory and Beyond" (World Scientific, Singapore, 1987)
5. Doniach S., Physica **B91**, 231 (1977)
6. Cannella V. and Mydosh J.A., Phys. Rev. **B6**, 4220 (1972)
7. Edwards S.F. and Anderson P.W., J. Phys. **F5**, 965 (1975)
8. Sherrington D. and Kirkpatrick S., Phys. Rev. Lett. **35**, 1792 (1975)
9. Sherrington D. and Southern B.W., J. Phys. **F5**, L49 (1975)
10. de Almeida J. and Thouless D.J., J. Phys. **A11**, 983 (1978)
11. Parisi G., J. Phys. **A13**, 1101 (1980)
12. Parisi G., Phys. Rev. Lett. **50**, 1946 (1983)
13. Nagata S., Keesom P.H. and Harrison H.R., Phys. Rev. **B19**, 1633 (1979)
14. Tholence J.L. and Tournier R., J. Phys. **35**, C4-229 (1974)
15. Mézard M., Parisi G., Sourlas N., Toulouse G. and Virasoro M.A., J. Phys. **45**, 843 (1984)

16. Mézard M. and Virasoro M.A., J. Phys. **46**, 1293 (1985)
17. Parisi G., in "Stealing the Gold: a Celebration of the Pioneering Physics of Sam Edwards", eds. P.M. Goldbart, N. Goldenfeld and D. Sherrington (OUP, Oxford, 2004), p. 192.
18. Viana L. and Bray A.J., J. Phys. **C18**, 3037 (1985)
19. Banavar J.R., Sherrington D. and Sourlas N., J. Phys. **A20**, L1 (1987)
20. Wong K.Y.M. and Sherrington D., J. Phys. **A21**, L359 (1988)
21. Sherrington D. and Mihill K., J. Phys. **35 C4**, 199 (1972)
22. Coles B.R., Tari A. and Jamieson H.C., in "Proc. Low Temp. Phys. LT-13", vol.2, (Plenum, New York 1974), p. 414
23. Anderson P.W., Phys. Rev. **109**, 1492 (1958)
24. Hertz J.A., Phys. Rev. **B19**, 4796 (1979)
25. Thouless D.J., Anderson P.W. and Palmer R.G., Phil. Mag. **35**, 137 (1977)
26. Ghatak S.K. and Sherrington D., J. Phys. **C10**, 3149 (1977)
27. Crisanti A. and Leuzzi L., Phys. Rev. Lett. **89**, 237204 (2002)
28. Rosenow B. and Oppermann R., Phys. Rev. Lett. **77**, 1608 (1996)
29. Pérez Castillo I. and Sherrington D., **B72**, 104427 (2005)
30. Sherrington D., The infinite-ranged m-vector spin glass in "Heidelberg Colloquium on Spin Glasses" eds. I. Morgenstern and L. van Hemmen (Springer, Berlin, Heidelberg, New York 1983), p. 125
31. Gabay M. and Toulouse G., Phys. Rev. Lett. **47**, 201 (1981)
32. Cragg D., Sherrington D. and Gabay M., Phys. Rev. Lett. **49**, 158 (1982)
33. Elderfield D. and Sherrington D., J. Phys. **A15**, L513 (1982)
34. Elderfield D. and Sherrington D., J. Phys. **C17**, 1923 (1984)
35. Cragg D. and Sherrington D., Phys. Rev. Lett. **49**, 1190 (1982)
36. Crisanti A. and Sommers H-J., Z. Phys. **B87**, 341 (1992)
37. Gross D., Kanter I. and Sompolinsky H., Phys. Rev. Lett. **55**, 304 (1985)
38. Goldbart P.M. and Sherrington D., J. Phys. **C18**, 1923 (1985)
39. Elderfield D. and Sherrington D., J. Phys. **C16**, L497 (1983)
40. Sherrington D., Prog. Theor. Phys. Suppl. **87**, 180 (1986)
41. Gardner E., Nuc. Phys. **B257**, 747 (1985)
42. Gillin P., Nishimori H. and Sherington D., J. Phys. **A34**, 2949 (2001)
43. Sherrington D., Neural networks: the spin glass approach in "Mathematical Studies of Neural networks", ed. J.G. Taylor (Elsevier, Amsterdam, 1992), p. 261
44. Nishimori H. "Statistical Physics of Spin Glasses and Information Processing"(OUP, Oxford, 2001)
45. de Dominicis C., Phys. Rev. **B18**, 4913 (1978)
46. Coolen A.C.C., Laugton S.N. and Sherrington D., Phys. Rev. **B53**, 8184 (1996)
47. Cugliandolo L. and Kurchan J., Phys. Rev. Lett. **71**, 173 (1993)
48. Cugliandolo L. and Kurchan J., J. Phys. **A26**, 5749 (1994)
49. Challet D., Marsili M. and Zhang Y-C., "Minority Games(OUP, Oxford, 2004)
50. Coolen A.C.C., "The Mathematical Theory of Minority Games" (OUP, Oxford, 2004)
51. Sherrington D., J. Phys. **C4**, 401 (1971)

Mean Field Models for Spin Glasses:
Some Obnoxious Problems

Michel Talagrand

Equipe d'analyse de l'institut Mathématique, Université Paris VI
Boite 186, 4 Place Jussieu, 75230 Paris Cedex 05, France
Department of Mathematics, The Ohio State University
100 Math Tower, 231 West 18th Avenue, Columbus, OH 43210-1174, USA
e-mail: talagran@math.ohio-state.edu

Summary. Despite steady progress over the last 10 years, the study of (the statics of) mean field models for spin glasses remains very difficult. We discuss some of the ideas behind the recent progresses, and some of the most blatant open problems.

1 Introduction

By many respects a (Ising mean field) spin glass model is a suitable random function U_N on $\Sigma_N = \{-1, 1\}^N$. The objective is the study (for large N) of the Gibbs measure, a random probability on Σ_N with density proportional to $\exp(U_N(\boldsymbol{\sigma}) + h \sum_{i \leq N} \sigma_i)$ where $\boldsymbol{\sigma} = (\sigma_i)_{i \leq N}$ denotes the generic element of Σ_N. Here, h is a given number (that represents the strength of an "external field") and, to lighten notation, we omit the negative signs that are customary in physics. Often U_N depends on a parameter β through the relation $U_N = \beta V_N$, and the problem is typically easier when β is small (or, more generally, under some kind of smallness condition on U_N).

These models are believed to have a very complex behavior at low enough temperature (large β), behavior that becomes (much) simpler at high temperature (small β). The appeal of the whole area (in my eyes at least) largely stems from the fact that this behavior apparently follows some universal laws (i.e., that are valid in all the models).

Roughly speaking, there are four stages of understanding of a given model:

(0) Inability to say anything at all
(1) Control at high enough temperature (i.e., small enough β)
(2) Control of the entire high temperature region
(3) Control at all temperature (all β)

Of course each of these stages (except maybe the first one) can be subdivided in substages describing at which level of accuracy the control is obtained.

Most of the recent book [1] is devoted to reaching the stage (1) for the main models. The main tool is the cavity method, that is, induction over the number N of coordinates. Probably it would have been better to coin another name, because what a mathematician means by cavity method is very different from what a physicist means.

What physicists do is:

(i) Assume that the overlaps are asymptotically equal to a constant q (see precise definition in (1.2)).

(ii) Compute q by establishing heuristically a self-consistent equation.

Part (i) requires no work, and part (ii) requires only little. Hence, for a physicist, the equation

$$\text{Cavity method} = \text{triviality} \tag{1.1}$$

holds true. The rational behind (1.) is that it is the "default situation." When asked why this is the case, Mézard answered: But what else could happen? The more I think about this argument, the more I find it convincing. Yet, it is not fully mathematical.

Equation (1.1) starts to be in doubt as soon as one tries to prove (i). Let us denote by $R_{1,2} = R_{1,2}(\boldsymbol{\sigma}^1, \boldsymbol{\sigma}^2) = N^{-1} \sum_{i \leq N} \sigma_i^1 \sigma_i^2$ the overlap of two configurations, by $\langle \cdot \rangle$ an average for the Gibbs measure (or its product on Σ_N^2) and E expectation in the disorder (i.e., the randomness of the Hamiltonian U_N). Then (i) means that

$$\lim_{N \to \infty} E \langle (R_{1,2} - q)^2 \rangle = 0. \tag{1.2}$$

The basic idea to prove this is to show that

$$E \langle (R_{1,2} - q)^2 \rangle \leq AE \langle (R_{1,2} - q)^2 \rangle + c_N \tag{1.3}$$

where $c_N \to 0$ and $A < 1$ does not depend on N. One starts by writing, using symmetry between sites

$$E \langle (R_{1,2} - q)^2 \rangle = E \langle (\sigma_N^1 \sigma_N^2 - q)(R_{1,2} - q) \rangle$$

and then one expresses the bracket as a function of the $(N-1)$-spin system. When conducted skillfully this computation yields the relation (1.3). The high-temperature condition greatly helps in obtaining the condition $A < 1$.

It does not seem possible however to prove directly (1.3) in all the region (1.2). As is explained in detail in Sect. 2.6 of [1], to perform this computation it seems necessary (unless one is at very high temperature) to know that

$$\text{it is very rare that } |R_{1,2} - q| \geq \varepsilon, \tag{1.4}$$

where $\varepsilon > 0$ is a sufficiently small number (the closer one is to the boundary of the high-temperature region, the smaller ε has to be). There seems to be no way to obtain (1.4) from the cavity method. In other words, while the cavity method achieves (1), it seems powerless to achieve (2) by itself. How then could one prove (1.4)? It would be satisfactory to prove a statement such as

$$E\langle 1_{\{|R_{1,2}-q|\geq\varepsilon\}}\rangle \leq K\exp\left(-\frac{N}{K}\right), \tag{1.5}$$

where K is independent of N. Now

$$\langle 1_{\{|R_{1,2}-q|\geq\varepsilon\}}\rangle = \frac{B_N(\varepsilon)}{Z_N^2}$$

where

$$B_N(\varepsilon) = \sum_{|R_{1,2}-q|\geq\varepsilon} \exp(H_N(\boldsymbol{\sigma}^1)+H_N(\boldsymbol{\sigma}^2))$$

$$Z_N = \sum \exp H_N(\boldsymbol{\sigma})$$

and

$$H_N(\boldsymbol{\sigma}) = U_N(\boldsymbol{\sigma}) + h\sum_{i\leq N}\sigma_i.$$

An extremely important consequence of concentration of measure is that to prove (1.5) it suffices to prove that

$$E\log B_N(\varepsilon) \leq 2E\log Z_N - \varepsilon'N, \tag{1.6}$$

where $\varepsilon' > 0$ is independent of N (see [1], Sect. 2.2). To prove this, it would be very nice to have an a priori upper bound on $E\log B_N(\varepsilon)$, and in particular a bound of the type

$$\frac{1}{N}E\log B_N(\varepsilon) \leq 2p - \frac{\varepsilon^2}{K_0}, \tag{1.7}$$

where K_0 is a number, independent of N and ε, and where we believe that

$$p = \lim_{N\to\infty} N^{-1}\log Z_N. \tag{1.8}$$

We then see the following scheme of proof: If we know (for large N) that

$$\frac{1}{N}E\log Z_N \geq p - \frac{\varepsilon^3}{3K_0} \tag{1.9}$$

then combining with (1.7) we get (1.6), hence (1.5), and thus (1.3), and this implies (1.2). Hopefully, once this strong information has been obtained, one can prove (1.8), from which we will be able to deduce (1.9) at a slightly lower temperature, and control in this manner the entire high-temperature region.

2 Guerra's Replica-Symmetric Bound

Before figuring out how to prove (1.7), one should study the simple problem of finding a bound for $p_N = N^{-1} E \log Z_N$. In the case of the Sherrington–Kirkpatrick (SK) model, F. Guerra has found a striking argument [2]. The Hamiltonian of the SK model is given by

$$H_N(\boldsymbol{\sigma}) = \frac{\beta}{\sqrt{N}} \sum_{i<j} g_{ij} \sigma_i \sigma_j + h \sum_{i \le N} \sigma_i, \tag{2.1}$$

where (g_{ij}) are i.i.d. standard Gaussian. Consider another independent sequence g_i of standard Gaussian r.v., that is independent of the previous sequence, and a parameter q. Guerra introduces the interpolating Hamiltonian

$$H_{N,t}(\boldsymbol{\sigma}) = \beta \left(\frac{\sqrt{t}}{\sqrt{N}} \sum_{i<j} g_{ij} \sigma_i \sigma_j + \sqrt{1-t} \sum_{i \le N} g_i \sqrt{q} \sigma_i \right) + h \sum_{i \le N} \sigma_i. \tag{2.2}$$

One key idea there is that for $t = 0$ there is no interaction between the sites, so the model is trivial. We define, for $0 \le t \le 1$,

$$\varphi_N(t) = \frac{1}{N} E \log \sum_{\sigma} \exp H_{N,t}(\boldsymbol{\sigma}). \tag{2.3}$$

Thus

$$\varphi_N(0) = \log 2 + E \log \operatorname{ch}(\beta g \sqrt{q} + h),$$

where g is standard Gaussian. Let us denote by $\langle \cdot \rangle_t$ an average with respect to the Gibbs measure with Hamiltonian (2.2). Then using integration by parts (see e.g. Theorem 2.4.7 of [1] for a detailed explanation) one sees that

$$\begin{aligned}
\varphi_N'(t) &= \frac{\beta^2}{2} \left(\frac{1}{2} E \langle R_{1,2}^2 \rangle_t - q(1 - E \langle R_{1,2} \rangle_t) \right) \\
&= \frac{\beta^2}{4} (1 - q)^2 - \frac{\beta^2}{4} E \langle (R_{1,2} - q)^2 \rangle_t \\
&\le \frac{\beta^2}{4} (1 - q)^2.
\end{aligned} \tag{2.4}$$

Consequently,

$$\begin{aligned}
p_N &= \frac{1}{N} E \log \sum_{\sigma} \exp H_N(\boldsymbol{\sigma}) = \varphi_N(1) \\
&\le RS(q) := \log 2 + \frac{\beta^2}{4} (1 - q)^2 + E \log \operatorname{ch}(\beta g \sqrt{q} + h).
\end{aligned} \tag{2.5}$$

The beauty of (2.5) is that it is obtained without any knowledge of the system. Since this inequality is valid for any q, we have

$$p_N \leq RS := \inf_q RS(q). \tag{2.6}$$

The importance of this bound is that it is tight in the high-temperature region (a statement that can be used as a definition of the high-temperature region).

This argument can be considerably generalized. Consider the case when

$$H_N(\boldsymbol{\sigma}) = U_N(\boldsymbol{\sigma}) + h \sum_{i \leq N} \sigma_i, \tag{2.7}$$

where $(U_N(\boldsymbol{\sigma}))$ is a jointly Gaussian family of r.v. such that for all configurations $\boldsymbol{\sigma}^1, \boldsymbol{\sigma}^2$, we have

$$\left| \frac{1}{N} E U_N(\boldsymbol{\sigma}^1) U_N(\boldsymbol{\sigma}^2) - \xi(R_{1,2}) \right| \leq c(N), \tag{2.8}$$

where $c(N) \to 0$ and ξ is a given function. Then if we now take

$$H_{N,t}(\boldsymbol{\sigma}) = \sqrt{t} U_N(\boldsymbol{\sigma}) + \sqrt{1-t} \sum_{i \leq N} g_i \sqrt{\xi'(q)} \sigma_i + h \sum_{i \leq N} \sigma_i, \tag{2.9}$$

where again the r.v. $(g_i)_{i \leq N}$ are i.i.d. standard Gaussian independent of U_N. Defining $\varphi_N(t)$ as before, we find that

$$\varphi'_N(t) = \frac{1}{2} \left(E \left\langle \frac{1}{N} E(U_N(\boldsymbol{\sigma})^2) \right\rangle_t - E \left\langle \frac{1}{N} E(U_N(\boldsymbol{\sigma}^1) U_N(\boldsymbol{\sigma}^2)) \right\rangle_t \right)$$

$$- \xi'(q)(1 - \langle R_{1,2} \rangle_t) \leq \frac{1}{2} \Big(\xi(1) - E \langle \xi(R_{1,2}) \rangle_t$$

$$- \xi'(q)(1 - E \langle R_{1,2} \rangle_t) \Big) + c(N)$$

using (2.9). If we assume that ξ is *convex* then

$$-\xi(R_{1,2}) + \xi'(q) R_{1,2} \leq \xi'(q) q - \xi(q)$$

and thus

$$\varphi'_N(t) \leq \frac{1}{2} (\xi(1) - \xi(q) + (q-1)\xi'(q)) + c(N)$$

and hence

$$\limsup_{N \to \infty} \frac{1}{N} E \log \sum_{\boldsymbol{\sigma}} \exp H_N(\boldsymbol{\sigma})$$

$$\leq \inf_q \left[\log 2 + \frac{1}{2} (\xi(1) - \xi(q) + (q-1)\xi'(q)) + E \log \mathrm{ch}(g \sqrt{\xi'(q)} + h) \right]. \tag{2.10}$$

The assumption (2.6) covers the case of the p-spin interaction model, where

$$V_N(\boldsymbol{\sigma}) = V_{N,p,\beta}(\boldsymbol{\sigma}) = \beta\left(\frac{p!}{2N^{p-1}}\right)^{1/2} \sum_{i_1 < \cdots < i_p} g_{i_1 \ldots i_p} \sigma_{i_1} \cdots \sigma_{i_p},$$

where of course $g_{i_1 \ldots i_p}$ are i.i.d. standard Gaussian. In this case $\xi(q) = \beta^2 q p/2$ is convex for p *even*. The function ξ is also convex in the more general case

$$V_N(\boldsymbol{\sigma}) = \sum_{p \text{ even}} V_{N,p,\beta_p}(\boldsymbol{\sigma}), \tag{2.11}$$

where the randomness of the terms in the sum are assumed to be independent.

It is in fact possible to extend (2.10) to the case where ξ is convex only on $[0,1]$ (so that all values of p become permitted in (2.11)) but this is rather nontrivial [3].

Here is an amusing and unimportant question: can one describe the class of functions ξ for which one can find functions U_N as in (2.9)?

An essential feature of the present scheme of proof is that the competent bound for p_N is the type $RS = \inf_q RS(q)$. Unfortunately, there are models of importance (the prime examples being the Perceptron model and the Hopfield model) for which the competent bound is of the type $\sup_r \inf_q RS(r, q)$ (or so it seems to me). Is there any hope in this case? To be specific, we will state only the simplest case we can think about (taking the risk that it is not a good choice), a kind of "Gaussian Hopfield" model. We consider independent standard Gaussian r.v. $(g_{ik})_{i,k \geq 1}$, and we set $S_{k,N}(\boldsymbol{\sigma}) = N^{-1/2} \sum_{i \leq N} g_{ik} \sigma_i$. Consider a parameter $\alpha > 0$, and $\beta < 1$.

Problem 2.1 *Find competent upper bounds for*

$$\limsup_{N \to \infty} \frac{1}{N} E \log \sum_{\sigma} \exp\left(\frac{\beta}{2} \sum_{k \leq \alpha N} S_{k,N}^2(\boldsymbol{\sigma}) + h \sum_{i \leq N} \sigma_i\right).$$

For small α (how small depending on β) there is little doubt that the methods of [1], Chap. 3, allow one to compute this limit. The problem is for the large values of α.

Let us now go back to our original problem, the search of a proof for (1.7). If one keeps in mind the probabilistic bound

$$E(e^X 1_{\{Y \geq a\}}) \leq e^{-\lambda a} E(e^{X + \lambda Y})$$

one will probably first write

$$N^{-1} E \log \sum_{R_{1,2} \geq u + \varepsilon} \exp(H_N(\boldsymbol{\sigma}^1) + H_N(\boldsymbol{\sigma}^2)) \leq$$

$$-\lambda(u + \varepsilon) + \frac{1}{N} E \log \sum_{\sigma^1, \sigma^2} \exp(H_N(\boldsymbol{\sigma}^1) + H_N(\boldsymbol{\sigma}^2) + N\lambda R_{1,2}),$$

and attempt to bound the last term. This is a problem of the same nature as bounding the quantity p_N of (2.5), except that now we have two copies of the system, copies that are coupled by the interaction term $\lambda N R_{1,2} = \lambda \sum_{i \leq N} \sigma_i^1 \sigma_i^2$. (This is a simple instance of the SK model with d-component spins as studied in [1], Sect. 12.13.)

When trying to extend the scheme of proof of (2.5) to this situation, one soon realizes that the fact that there are two copies of the model creates an interaction term that has the wrong sign to make the argument work. On the other hand, the fact that we deal with coupled copies (as opposed to uncoupled copies) means that through general principles (use of convexity as explained in [1], Sect. 2.12), for most values of λ one can expect $E \langle (R_{1,2} - E \langle R_{1,2} \rangle)^2 \rangle$ to be small, and the terms with a wrong sign need not to be devastating. The paper [4] follows this approach, that unfortunately gets bogged into impossible difficulties (but that I mention as I still hope that it could be useful at some point).

The difficulty created by these terms with a wrong sign is bypassed by the simple device of restricting the coupled system to a set where $R_{1,2}$ is constant (a drastic way to control the value of $R_{1,2}$), i.e., to control

$$\frac{1}{N} E \log \sum_{R_{1,2}=u} \exp(H_N(\boldsymbol{\sigma}^1) + H_N(\boldsymbol{\sigma}^2)). \tag{2.12}$$

Controlling this for a single value of u with $|u - q| \geq \varepsilon$ is almost as good as controlling the similar quantity where the summation is over $|R_{1,2} - q| \geq \varepsilon$, because there are at most $2N + 1$ possible values of u, since $N R_{1,2}$ is an integer. To control the quantity (2.12), consider a pair of jointly Gaussian r.v. (g^1, g^2) with $E(g^1)^2 = E(g^2)^2 = q$, $Eg^1g^2 = \varrho$, and (g_i^1, g_i^2) independent copies of this pair, that are also independent of the r.v. (g_{ij}) of (2.1). For $0 < t < 1$, consider

$$H_{N,t}(\boldsymbol{\sigma}^1, \boldsymbol{\sigma}^2) = \beta \left(\sqrt{\frac{t}{N}} \sum_{i<j} g_{ij}(\sigma_i^1 \sigma_j^1 + \sigma_i^2 \sigma_j^2) + \sqrt{1-t} \sum_{i \leq N} (g_i^1 \sigma_i^1 + g_i^2 \sigma_i^2) \right)$$
$$+ h \sum_{i \leq N} (\sigma_i^1 + \sigma_i^2),$$

and

$$\varphi_N(t) = \frac{1}{N} E \log \sum_{R_{1,2}=u} \exp H_{N,t}(\boldsymbol{\sigma}^1, \boldsymbol{\sigma}^2).$$

Proceeding as in (2.4) we obtain

$$\varphi_N'(t) \leq \frac{\beta^2}{2}((1 - q)^2 + (u - \varrho)^2). \tag{2.13}$$

Now, for $\lambda \geq 0$, we have

$$\varphi_N(0) = \frac{1}{N} E \log \sum_{R_{1,2}=u} \exp\left(\sum_{i \leq N} g_i^1 \sigma_i^1 + g_i^2 \sigma_i^2\right)$$

$$= -\lambda u + \frac{1}{N} \log \sum_{R_{1,2}=u} \exp\left(\sum_{i \leq N}(g_i^1 \sigma_i^1 + g_i^2 \sigma_i^2) + \lambda \sum_{i \leq N} \sigma_i^1 \sigma_i^2\right)$$

$$\leq -\lambda u + \frac{1}{N} \log \sum_{\sigma^1, \sigma^2} \exp\left(\sum_{i \leq N}(g_i^1 \sigma_i^1 + g_i^2 \sigma_i^2) + \lambda \sum_{i \leq N} \sigma_i^1 \sigma_i^2\right)$$

$$= -\lambda u + 2\log 2 + E \log(\mathrm{ch}(g_1+h)\mathrm{ch}(g_2+h)\mathrm{ch}\lambda + \mathrm{sh}(g_1+h)\mathrm{sh}$$
$$\times (g_2+h)\mathrm{sh}\lambda)$$

and we get the bound

$$\frac{1}{N} E \log \sum_{R_{1,2}=u} \exp(H_N(\sigma^1) + H_N(\sigma^2))$$

$$\leq 2\log 2 + \inf\left(-\lambda u + \frac{\beta^2}{2}((1-q)^2 + (u-\varrho)^2)\right.$$

$$+ E \log\left(\mathrm{ch}(\beta g_1 + h)\mathrm{ch}(\beta g_2 + h)\mathrm{ch}\lambda\right.$$

$$\left.\left. + \mathrm{sh}(\beta g_1 + h)\mathrm{sh}(\beta g_2 + h)\mathrm{sh}\lambda\right)\right), \qquad (2.14)$$

where the infimum is taken over all possible choices of q and ϱ. (The case $q = \varrho$ of this result is considered in Proposition 2.9.6 of [1].)

Let us now make a (very unpleasant) observation, that introduces us to one of the recurrent ideas of this paper. For any u, the left-hand side of (2.14) is at most $2p_N$ (where p_N is defined in (2.5)). Yet, it does not seem obvious at all that the right-hand side of (2.14) is less than twice the right-hand side of (2.6). It is not difficult to prove this when β is small enough (which is fortunate since in that case the bound (2.14) becomes tight as $N \to \infty$). The problem is for large β.

The discrepancy between the bounds (2.6) and (2.14) has a simple explanation. The bound (2.14) can be improved for large β, as we will show in Sect. 3. Later in the paper we will find discrepancies between bounds for which we have yet no explanation to offer.

3 Guerra's Broken Replica-Symmetry Bound

One can argue that after all the bound (2.6) relies on the standard technique of Gaussian interpolation (and on a serious dose of optimism, just to believe that such a miracle is possible). The far-reaching generalization presented here (also due to Guerra) must have required a far deeper insight. We will directly explain this procedure in the general setting of (2.7) and (2.9). We consider an integer k and numbers $0 = m_0 < m_1 < \cdots \le m_k = 1$, and $0 = q_0 \le q_1 \le \cdots \le q_{k+1} = 1$. We write

$$\boldsymbol{m} = (m_0, \ldots, m_k); \qquad \boldsymbol{q} = (q_0, \ldots, q_k, q_{k+1}). \qquad (3.1)$$

Consider independent Gaussian r.v. $(z_p)_{0 \le p \le k}$ with

$$E z_p^2 = \xi'(q_{p+1}) - \xi'(q_p),$$

and independent copies $(z_{i,p})_{i \le N}$ of the sequence $(z_p)_{0 \le p \le k}$, that are of course independent of the randomness of H_N. For $0 \le t \le 1$ we consider the Hamiltonian

$$H_{N,t}(\boldsymbol{\sigma}) = \sqrt{t} H_N(\boldsymbol{\sigma}) + \sqrt{1-t} \sum_{i \le N} \sigma_i \Big(\sum_{0 \le p \le k} z_{i,p} + h \Big).$$

We define (keeping the dependence in N implicit)

$$F_{k+1,t} = \log \sum_{\sigma} \exp H_{N,t}(\boldsymbol{\sigma}).$$

Denoting by E_l expectation in the r.v. $(z_{l,p})$ for $i \le N$ and $p \ge l$, we define recursively

$$F_{l+1,t} = \frac{1}{m_l} \log E_l \exp m_l F_{l+1,t}.$$

When $m_l = 0$ this means that $F_{l,t} = E_l F_{l+1,t}$. We set

$$\varphi_N(t) = \frac{1}{N} E F_{1,t},$$

where the expectation is now in the randomness of H_N and the variables $z_{i,0}$. We define

$$W_l = \exp m_l (F_{l+1,t} - F_{l,t}).$$

Denoting by $\langle \cdot \rangle_t$ an average for the Gibbs measure with Hamiltonian H_t it is easily seen that the map

$$f \mapsto E_l(W_l \cdots W_k \langle f \rangle_t)$$

is given by integration against a probability measure γ_l on $\Sigma_N = \{-1, 1\}^N$. We denote by $\gamma_l^{\otimes 2}$ its product on $\Sigma_N^{\otimes 2}$, and for a function f on Σ_N^2 we set

$$\mu_l(f) = E(W_1 \cdots W_{l-1} \gamma_l^{\otimes 2}(f)).$$

We define

$$\theta(q) = q\xi'(q) - \xi(q). \qquad (3.2)$$

Theorem 3.1. (Guerra's identity [5]) *We have*

$$\varphi_N'(t) = -\frac{1}{2} \sum_{1 \leq l \leq k} m_l(\theta(q_{l+1}) - \theta(q_l))$$

$$- \frac{1}{2} \sum_{1 \leq l \leq k} (m_l - m_{l-1})\mu_l(\xi(R_{1,2}) - R_{1,2}\xi'(q_l) + \theta(q_l)) + \mathcal{R}, \tag{3.3}$$

where $|\mathcal{R}| \leq c(N)$.

When ξ is *convex*, we have $\xi(x) - x\xi'(q) + \theta(q) \geq 0$, so that (3.3) yields

$$\varphi_N'(1) \leq \varphi(0) - \frac{1}{2} \sum_{1 \leq l \leq k} m_l(\theta(q_{l+1}) - \theta(q_l)) + c(N). \tag{3.4}$$

To compute $\varphi(0)$, we consider the r.v. $X_{k+1} = \log \mathrm{ch}(h + \sum_{0 \leq p \leq k} z_p)$, and, recursively, for $l \geq 1$ we define $X_l = m_l^{-1} \log E_l \exp m_l X_{l+1}$, where E_l denotes expectation in the r.v. z_p, $p \geq l$. When $m_l = 0$, this means that $X_l = E_l X_{l+1}$. We define

$$X_0 = E X_1, \tag{3.5}$$

and it is very easy to check that $\varphi(0) = \log 2 + X_0$. We define

$$P(k, q, m) = \log 2 + X_0 - \frac{1}{2} \sum_{1 \leq l \leq k} m_l(\theta(q_{l+1}) - \theta(q_l)), \tag{3.6}$$

so that (3.4) implies

$$\varphi_N(1) \leq P(k, q, m) + c(N).$$

Since this bound holds for all the choices of k, q, and m, we have the following.

Theorem 3.2. (Guerra's bound) *If ξ is convex, we have*

$$\limsup_{N \to \infty} \frac{1}{N} E \log \sum_{\sigma} \exp H_N(\sigma) \leq P := \inf P(k, q, m), \tag{3.7}$$

where the infimum is over all possible choices of the parameters.

Of course the letter P stands for Parisi, who invented this expression. It was recently shown that in some cases the bound (3.7) is tight, as the following explains.

Theorem 3.3. [10] *Assume that ξ is convex, that $\xi(x) = \xi(-x)$, $\xi(0) = 0$ and $\xi''(x) > 0$ for $x > 0$. Given $t_0 < 1$ we can find $\varepsilon > 0$ with the following property. Consider k, q and m such that*

$$P(k, q, m) \leq P + \varepsilon \tag{3.8}$$

$$P(k, q, m) \quad \text{is minimum over all choices of } q \text{ and } m \tag{3.9}$$

as in (3.1). Then for $t \leq t_0$ we have

$$\lim_{N \to \infty} \varphi_N(t) = \log 2 + X_0 - \frac{t}{2} \sum_{0 \leq l \leq k} m_l(\theta(q_{l+1}) - \theta(q_l)), \tag{3.10}$$

where X_0 is as in (3.5). Consequently, we have

$$\lim_{N \to \infty} \frac{1}{N} E \log \sum_{\sigma} \exp H_N(\boldsymbol{\sigma}) = \mathcal{P}.$$

While there is a technical element in the proof of (3.10), due to the need of using condition (3.9), the basic idea is somehow an elaboration of the ideas explained in Sect. 1 (although the cavity method is not used). One proves that the remainder terms in Guerra's identity (3.3) are small, and the main tools are a priori bounds in the spirit of (2.14). These bounds involve the functionals μ_l. For simplicity we will discuss here only bounds involving the Gibbs measure. These bounds extend (2.14) the way (3.7) extends (2.10). We consider an integer $\tau \geq 1$, and numbers $0 = n_0 < \cdots \leq n_\tau = 1$. For $0 \leq p \leq \tau$, we consider independent pairs (z_p^1, z_p^2) of jointly Gaussian r.v. For $j, j' \in \{1, 2\}$, $l \leq \tau$, we define the numbers $q_l^{j,j'}$ by

$$\xi'(q_l^{j,j'}) = \sum_{0 \leq p \leq l} E z_p^j z_p^{j'}. \tag{3.11}$$

We assume that $q_\tau^{1,1} = q_\tau^{2,2} = 1$. Given $\lambda > 0$, we define the r.v.

$$Y_{\tau+1,\lambda} = \log \Big(\text{ch}\Big(h + \sum_{p \leq \tau} z_p^1 \Big) \text{ch}\Big(h + \sum_{p \leq \tau} z_p^2 \Big) \text{ch}\lambda$$

$$+ \text{sh}\Big(h + \sum_{p \leq \tau} z_p^1 \Big) \text{sh}\Big(h + \sum_{p \leq \tau} z_p^2 \Big) \text{sh}\lambda \Big)$$

and, recursively,

$$Y_{l,\lambda} = \frac{1}{n_l} \log E_l \exp n_l Y_{l+1,\lambda},$$

where E_l denotes expectation in the r.v. z_p^j, $p \geq l$.

Proposition 3.4. *For all choices of parameters as above we have*

$$\frac{1}{N} E \log \sum_{R_{1,2}=u} \exp(H_N(\boldsymbol{\sigma}^1) + H_N(\boldsymbol{\sigma}^2))$$

$$\leq 2 \log 2 + Y_{0,\lambda} - \lambda u - \frac{1}{2} \sum_{j,j' \leq 2} \sum_{1 \leq l \leq \tau} n_l(\theta(q_{l+1}^{j,j'}) - \theta(q_l^{j,j'})) \tag{3.12}$$

$$+ \xi(u) - \xi(q_{\tau+1}^{1,2}) - (u - q_{\tau+1}^{1,2})\xi'(q_{\tau+1}^{1,2}) + 4c(N).$$

To prove the bound (2.14), one takes (z_0^1, z_0^2) such that $E((z_0^1)^2) = E((z_1^2)^2) = \beta^2 q$, $E(z_0^1 z_0^2) = \beta^2 \varrho$ and (z_1^1, z_1^2) independent of variance $\beta^2(1-q)$.

In the case of the SK model, Proposition 3.4 is essentially Theorem 2.11.14 of [1]. My intuition is that the infimum of the right-hand side of (3.12) over all choices of parameters should not increase if one adds the requirement that $u = q_{\tau+1}^{1,2}$. Still, the general formulation allowing $u \neq q_{\tau+1}^{1,2}$ could be useful.

A very important feature of the bound (3.12) is that it avoids the problems with the bound (2.14) that were discussed at the end of Sect. 2.2. Namely, the infimum over all choices of parameters is $\leq 2\mathcal{P}$, an observation that is in fact the main key to Theorem 3.3. To prove this, we consider k, $\boldsymbol{m}$, and $\boldsymbol{q}$ as in (3.1), and we choose the parameters in (3.12) to obtain a right-hand side that is equal to $2\mathcal{P}(k, \boldsymbol{q}, \boldsymbol{m})$. We consider only the case $u \geq 0$ (which is the important one). Without loss of generality we can assume that $u = q_p$ for some p (otherwise we insert u at its proper place in the sequence $\boldsymbol{q}$). We then take $\tau = n$, $n_l = m_l/2$ if $l < p$, $n_l = m_l$ if $l \geq p$. For $l < p$, we choose $z_l^1 = z_l^2$, with variance $\xi'(q_{l+1}) - \xi'(q_l)$. If $l \geq p$, we choose z_l^1 and z_l^2 independent, with variance $\xi'(q_{l+1}) - \xi'(q_l)$. When $\lambda = 0$, it is straightforward to see that the right-hand side of (3.12) is $2\mathcal{P}(k, \boldsymbol{q}, \boldsymbol{m})$. In fact, $Y_{0,0} = 2X_0$ and

$$\sum_{j,j=1,2} \sum_{1 \leq l \leq n} n_l(\theta(q_{l+1}^{j,j'}) - \theta(q_l^{j,j'})) = 2 \sum_{1 \leq l \leq n} m_l(\theta(q_{l+1}) - \theta(q_l)).$$

Given sequences $\boldsymbol{m}$ and $\boldsymbol{q}$ as in (3.1) it is convenient to think of m_l as a parameter attached to the interval $[q_l, q_{l+1}[$. The couple $(\boldsymbol{q}, \boldsymbol{m})$ then represents a pairwise constant nondecreasing function $x_{\boldsymbol{q},\boldsymbol{m}}$. Let us write $\mathcal{P}(x_{\boldsymbol{q},\boldsymbol{m}})$ rather than $\mathcal{P}(k, \boldsymbol{q}, \boldsymbol{m})$. It can be shown that the map $x_{\boldsymbol{q},\boldsymbol{m}} \mapsto \mathcal{P}(x_{\boldsymbol{q},\boldsymbol{m}})$ extends nicely to the class of non decreasing functions x with $x(1) = 1$. The extension is pointwise continuous (when $\xi(q) = \beta^2 q^2/2$, a nice quantitative version of this statement is already given in [5]). The pointwise continuity of the map $x \mapsto \mathcal{P}(x)$ implies that there exists x such that $\mathcal{P}(x)$ is minimum.

Problem 3.5 *Prove that the minimum is obtained at a unique point x.*

A probability measure ν on $[0, 1]$ such that

$$\forall q, \quad \nu([0, q]) = x(q)$$

while $\mathcal{P}(x) = \inf_y \mathcal{P}(y)$ will be called a *Parisi measure*. (So, there is always at least one Parisi measure, but we do not know about uniqueness.)

Let us go back to the case of the Hamiltonian (2.11), so that $\xi(q) = \sum \beta_p^2 q^p$, where the sum is over p even, $p \geq 2$. To avoid convergence problems we assume $\sum 4^p \beta_p < \infty$.

Theorem 3.6. [6] *Assume $h \neq 0$. If $\beta_p \neq 0$ for all p even, the Parisi measure ν is unique. For every continuous function f on $[0, 1]$, we have*

$$\lim_{N \to \infty} E\langle f(R_{1,2}) \rangle = \int f \, d\nu. \tag{3.13}$$

In words, ν is the limiting law of $R_{1,2}$. Of course one would like to prove this theorem for all sequences (β_p). The condition $h \neq 0$ is simply to avoid symmetry around zero.

4 The Ultrametricity Conjecture

One of the most famous predictions about spin glasses is ultrametricity. To state the conjecture, we consider three replicas, i.e., the product Gibbs measure on $\Sigma_N^3 = (\{-1,1\}^N)^3$, and for a point $(\boldsymbol{\sigma}^1, \boldsymbol{\sigma}^2, \boldsymbol{\sigma}^3)$ of this space we write $R_{j,j'} = N^{-1} \sum_{i \leq N} \sigma_i^j \sigma_i^{j'}$. The ultrametricity conjecture states that (when Σ_N^3 is weighted with Gibbs' measure) the Hamming distance $d(\boldsymbol{\sigma}^1, \boldsymbol{\sigma}^2) = 1 - R_{1,2}$ is ultrametric, i.e.,

$$d(\boldsymbol{\sigma}^1, \boldsymbol{\sigma}^3) \leq \max(d(\boldsymbol{\sigma}^1, \boldsymbol{\sigma}^2), d(\boldsymbol{\sigma}^2, \boldsymbol{\sigma}^3))$$

or, equivalently,

$$R_{1,3} \geq \min(R_{1,2} R_{2,3}).$$

The weak ultrametricity conjecture:

$$\forall \varepsilon > 0, \quad \lim_{N \to \infty} E\langle 1_{\{R_{1,3} \leq \min(R_{1,2}, R_{2,3}) - \varepsilon\}}\rangle = 0. \tag{4.1}$$

The strong ultrametricity conjecture:

$$\forall \varepsilon > 0, \quad E\langle 1_{\{R_{1,3} \leq \min(R_{1,2}, R_{2,3}) - \varepsilon\}}\rangle \leq K \exp\left(-\frac{N}{K}\right), \tag{4.2}$$

where K does not depend on N.

It has been understood (mainly by G. Parisi [7], see also [8]) that (3.13), together with ultrametricity and the (generalized) Ghirlanda–Guerra identities [9] would give a complete description of the joint limiting law of any finite number of overlaps $R_{j,j'}$ (a result that contains much information about the system). We do not want to discuss here the Ghirlanda–Guerra identities (see Sects. 3.4 and 6.4 of [1]), but we want to stress that, while the identities are true "in average" over the value of the parameters (e.g., the parameters β_p of the end of Sect. 3) it does not seem to me to be known how to prove that they hold as $N \to \infty$ for *any value whatsoever* of the parameters where low temperature behavior occurs. (And statements to the contrary that are in print should be dismissed.) This is related to an especially vexing question, that we explain now in a simple case, that of the SK model. Consider

$$p_N(\beta) = \frac{1}{N} E \log \sum \exp\left(\frac{\beta}{\sqrt{N}} \sum_{i<j} g_{ij} \sigma_i \sigma_j\right).$$

This function is convex, and $p_N'(\beta_0)$ remains bounded as $N \to \infty$ for any β (in fact, also independently of β). Thus $p_N''(\beta) \geq 0$ remains bounded "in average in any interval."

Problem 4.1 *Prove that for any value of β_0 one has*

$$\sup_N \sup_{\beta \leq \beta_0} p_N''(\beta) < \infty. \tag{4.3}$$

The following weaker version would also be of interest.

Problem 4.2 *Prove that for any value of β_0 one has*

$$\lim_{N \to \infty} \frac{1}{N} \sup_{\beta \leq \beta_0} p_N''(\beta) = 0. \tag{4.4}$$

In fact, even the following is open.

Problem 4.3 *Prove that there exists a value of $\beta > 1$ for which one has* (4.4).

It is not difficult to see that $|p_N''(\beta)| \leq LN$, where L does not depend on N or β, so that (4.4) is the weakest nontrivial statement. It is this kind of statement one would need to prove the Ghirlanda–Guerra identities at any given value of the parameters.

A natural method to attack the strong ultrametricity conjecture would be, given $u_{1,2}, u_{1,3}, u_{2,3}$, to find bounds for

$$\frac{1}{N} E \log \sum_{R_{j,j'}=u_{j,j'}} \exp(H_N(\sigma^1) + H_N(\sigma^2) + H_N(\sigma^3)), \tag{4.5}$$

where the summation is over all triples $\sigma^1, \sigma^2, \sigma^3$ with $R_{j,j'} = u_{j,j'}$ for $1 \leq j \leq j' \leq 3$, in such a way that these bounds stay away from $3P$ as $u_{1,3}$ stays below $\min(u_{1,2}, u_{2,3})$. There is a natural extension of Proposition 3.4 to the case of the expression (4.5). Consider an integer τ, and numbers $0 = n_0 \leq \cdots \leq n_{\tau+1} = 1$. For $0 \leq p \leq \tau$, consider independent triples (z_p^1, z_p^2, z_p^3) of jointly Gaussian r.v. For $1 \leq j, j' \leq 3$, define $q_l^{j,j'}$ by

$$\xi'(q_l^{j,j'}) = \sum_{0 \leq p \leq l} E(z_p^j z_p^{j'}).$$

Given numbers $\lambda_1, \lambda_2, \lambda_3$, consider the r.v.

$$
Y_{\tau+1,\lambda_1,\lambda_2,\lambda_3} = \log\Bigg(\sum_{\varepsilon_1,\varepsilon_2,\varepsilon_3=\pm 1} \exp\Bigg(\sum_{j \leq 3} \varepsilon_j \Big(h + \sum_{p \leq \tau} z_p^i \Big) \\
+ \lambda_1 \varepsilon_2 \varepsilon_3 + \lambda_2 \varepsilon_1 \varepsilon_3 + \lambda_3 \varepsilon_1 \varepsilon_2 \Bigg) \Bigg),
$$

and define the r.v. $Y_{l,\lambda_1,\lambda_2,\lambda_3}$ recursively as usual.

Proposition 4.4. *For choices of these parameters as above, we have*

$$\frac{1}{N} E \log \sum_{R_{j,j'}=u_{j,j'}} \exp\left(\sum_{j\leq 3} H_N(\boldsymbol{\sigma}^j)\right)$$

$$\leq \sum_{1\leq j<j'\leq 3} \xi(u_{j,j'}) - \xi(q_{\tau+1}^{j,j'}) - (u_{j,j'} - q_{\tau+1}^{j,j'})\xi'(q_{\tau+1}^{j,j'}) \qquad (4.6)$$

$$-\frac{1}{2} \sum_{j,j'=1,2,3} \sum_{1\leq l\leq\tau} n_l(\theta(q_{l+1}^{j,j'}) - \theta(q_l^{j,j'}))$$

$$+ Y_{0,\lambda_1,\lambda_2,\lambda_3} - \lambda_1 u_{2,3} - \lambda_2 u_{1,3} - \lambda_3 u_{1,2} + 9c(N).$$

The puzzle is as follows. Given the numbers $u_{1,2}$, $u_{1,3}$, $u_{2,3}$, it is absolutely unclear to me that the infimum of the right-hand side of (4.6) is $\leq 3\mathcal{P}$ (so we are back in a situation as unpleasant as (2.14)). The only case where this is clear is when the numbers $u_{j,j'}$ satisfy the ultrametricity condition. Let us explain what happens in the case $0 \leq u = u_{1,3} = u_{2,3} \leq u_{1,2} = v \leq 1$. Considering sequences $\boldsymbol{q}, \boldsymbol{m}$ as in (3.1), we want to find parameters showing that the right-hand side of (4.6) is at most $3\mathcal{P}(k,\boldsymbol{q},\boldsymbol{m})$. Without loss of generality we can assume that $u = q_r$, $v = q_s$. We consider the subset of $[0,1]$ that consists of the numbers $m_{l/3}$, $l < r$, the numbers m_l, $m_{l/2}$ for $r \leq l < s$, and the numbers m_l, $r < l \leq k$. For clarity we assume that it does not occur that $m_l = m_{l'/2}$ for $r \leq l, l' < s$. We enumerate these numbers as a nondecreasing sequence (n_l), $0 \leq l \leq \tau$. For $l < r$ we have $n_l = m_{l/3}$, and we take $z_l^1 = z_l^2 = z_l^3$, of variance $\xi'(q_{l+1}) - \xi'(q_l)$. If $n_l = m_p/2$ for $r \leq p < s$, $z_l^1 = z_l^2$ is of variance $\xi'(q_{p+1}) - \xi'(q_p)$, and $z_l^3 = 0$. If $n_l = m_p$ for $r \leq p < s$, then $z_l^1 = z_l^2 = 0$ and z_l^3 has variance $\xi'(q_{p+1}) - \xi'(q_l)$. Finally, if $n_l = m_p$ for $p > s$, then z_l^1, z_l^2, z_l^3 are independent, of variance $\xi'(q_{p+1}) - \xi'(q_p)$. We then take $\lambda_1 = \lambda_2 = \lambda_3 = 0$, and we check easily that the right-hand side of (4.6) is then equal to $3\mathcal{P}(k,\boldsymbol{q},\boldsymbol{m})$.

The situation in Proposition 4.4 is made complicated by the fact that we are dealing with triplets; dimension 3 is often much harder than dimension 2. As Sect. 5 shows, the same puzzle already arises in dimension 2.

5 The Chaos Problem

The general version of this problem is as follows [11–13]. We consider two random functions U_N^1, U_N^2 on Σ_N, that are jointly Gaussian (centered); that is, the family of r.v. $U_N^1(\boldsymbol{\sigma}^1), U_N^2(\boldsymbol{\sigma}^2)$ for all values of $\boldsymbol{\sigma}^1$ and $\boldsymbol{\sigma}^2$ is jointly Gaussian. Consider two real numbers h_1, h_2, and the Hamiltonian

$$H_N^j(\boldsymbol{\sigma}) = U_N^j(\boldsymbol{\sigma}) + h_j \sum_{i\leq N} \sigma_i.$$

Consider the Hamiltonian

$$H_N^1(\boldsymbol{\sigma}^1) + H_N^2(\boldsymbol{\sigma}^2)$$

on Σ_N^2, and the corresponding Gibbs measure. When is it true that the overlap $R_{1,2}$ takes only essentially one value? (In physics, this is interpreted by saying that the "states" corresponding to the two systems with Hamiltonian H_N^j are unrelated). In the case where $U_N^1 = U_N^2$, if there is chaos (the overlap $R_{1,2}$ takes essentially one value) as soon as $h_1 \neq h_2$, one calls the situation "chaos in external field." If $h_1 = h_2$ and $U_N^1 = \beta_1 V_N$, $U_N^2 = \beta_2 V_N$, if there is chaos as soon as $\beta_1 \neq \beta_2$, one calls the situation "chaos in temperature."

Let us assume that for some functions $\xi_{j,j'}$, for $j, j' = 1, 2$, we have

$$|\xi_{j,j'}(R_{1,2}) - \frac{1}{N}E(U_N^j(\boldsymbol{\sigma}^1)U_N^{j'}(\boldsymbol{\sigma}^2))| \leq c(N).$$

We define the functions

$$\theta_{j,j'}(q) = q\xi'_{j,j'}(q) - \xi_{j,j'}(q).$$

Consider numbers $0 = n_0 \leq \cdots \leq n_\tau = 1$ and for $0 \leq p \leq \tau$ consider independent pairs (z_p^1, z_p^2) of jointly Gaussian centered r.v. We define the numbers $q_l^{j,j'}$ by

$$\xi'_{j,j'}(q_l^{j,j'}) = \sum_{0 \leq p < l} Ez_p^j z_p^{j'}.$$

Consider a number λ. Define

$$X_{\tau+1,\lambda} = \log\left(\mathrm{ch}\left(h_1 + \sum_{0 \leq p \leq \tau} z_p^1\right) \mathrm{ch}\left(h_2 + \sum_{0 \leq p \leq \tau} z_p^2\right) \mathrm{ch}\lambda \right.$$

$$\left. + \mathrm{sh}\left(h_1 + \sum_{0 \leq p \leq \tau} z_p^1\right) \mathrm{sh}\left(h_2 + \sum_{0 \leq p \leq \tau} z_p^2\right) \mathrm{sh}\lambda \right)$$

and define recursively $X_{l,\lambda}$ as usual.

Proposition 5.1. *If the functions $\xi_{j,j'}$ are convex, we have*

$$\frac{1}{N}E\log\sum_{R_{1,2}=u} \exp(H_N^1(\boldsymbol{\sigma}^1) + H_N^2(\boldsymbol{\sigma}^2))$$

$$\leq \xi_{1,2}(u) - \xi_{1,2}(q_{\tau+1}^{1,2}) - (u - q_{\tau+1}^{1,2})\xi'_{1,2}(q_{\tau+1}^{1,2})$$

$$+ 2\log 2 + X_{0,\lambda} - \lambda u \qquad (5.1)$$

$$- \frac{1}{2}\sum_{j,j'=1,2}\sum_{1 \leq l \leq \tau} n_l(\theta_{j,j'}(q_{l+1}^{j,j'}) - \theta_{j,j'}(q_l^{j,j'})) + c(N).$$

The problem again is that, unless $u = 0$, it does not seem obvious at all that the infimum of the right-hand side over all values of parameters is $\leq \mathcal{P}_1 + \mathcal{P}_2$ (with obvious notation). I see only three possibilities:

(a) *The bound* (5.1) *is not the correct one.* Maybe I do not understand well the "general Parisi conjecture," but it seems to me that, under this conjecture, asymptotically the left-hand side of (5.1) is given by the infimum of the right-hand side over all parameters, even if one adds the constraint $q_{\tau+1}^{1,2} = u$. Would it be possible that the Parisi theory is incomplete?

(b) There is some kind of analytical structure behind an expression like the right-hand side of (5.1) that ensures that the infimum over the parameters is $\leq \mathcal{P}_1 + \mathcal{P}_2$ for a (very) nontrivial reason.
This, of course, is the most exciting possibility. The theory of spin glasses will probably influence Probability theory in a lasting manner. Will it also eventually influence analysis?

(c) *I am totally confused!* The chaos problem is of particular interest in the case $h_1 = h_2 = 0$, $U_N^1 = \beta_1 V_N$, $U_N^2 = \beta_2 V_N$, where

$$V_N(\boldsymbol{\sigma}) = \left(\frac{p!}{N^{p-1}}\right)^{1/2} \sum_{i_1 < \cdots < i_p} g_{i_1 \ldots i_p} \sigma_{i_1} \cdots \sigma_{i_p}$$

for some p given. In that case

$$\xi_{j,j'}(q) = \beta_j \beta_{j'} q^p; \qquad \theta_{j,j'}(q) = (p-1)\beta_j \beta_{j'} q^p.$$

When β_1 and β_2 are not too large, the value of $\mathcal{P}_j$ is obtained as $\mathcal{P}(k, \boldsymbol{q}, \boldsymbol{m})$ for $k = 3$, i.e., equal to the infimum over the choices of m and q of

$$\log 2 + \frac{\beta_j^2}{4}(1 - pq^{p-1}) + \frac{\beta_j^2}{4}(p-1)q^p(1-m) + \frac{1}{m} \log E\mathrm{ch}\left(\beta z \sqrt{pq^{p-1}}\right)$$

where z is standard normal.

One would certainly expect in that case the infimum of the right-hand side of (5.1) to be obtained for a small value of τ (probably $\tau \leq 4$). Let us denote by $\varphi(u)$ the value of this infimum. One expects that (as in the case $\beta_1 = \beta_2$) when u increases from 0, $\varphi(u)$ decreases at first, so we really have to care only for the values of u such that $\varphi(u)$ is a local maximum. In that case, if one assumes, as is probably true, that the infimum of the right-hand side of (5.1) is obtained for $q_{\tau+1}^{1,2} = u$, one sees that when $\varphi(u)$ is a local minimum, one should have $\lambda = 0$. In other words, we cannot really take advantage of the parameter λ in the bound (5.1), and the bound becomes even simpler, and the mystery even deeper.

Acknowledgement

This work was partially supported by an NSF grant.

References

1. Talagrand, M (2003) Spin Glasses, a challenge for Mathematicians, Springer, Berlin, Heidelberg, Newyork, 596 p.
2. Guerra, F (2001) *Sum rules for the free energy in the mean field spin glass model*, Field Inst. Commun. 30, 161
3. Talagrand, M. (2003) *The generalized Parisi formula*, C. R. Acad. Sci. Paris 307, 2003, no. 2, pp. 111–114
4. Talagrand, M. (2002) *On the high-temperature phase of the Sherrington–Kirkpatrick model*, Ann. Probab. 30, pp. 364–481
5. Guerra, F. (2003) *Replica broken bounds in the mean field spin glass model*, Commun. Math. Phys. 233, pp. 1–12
6. Talagrand, M. (2006) *Parisi measures*, J. Func. Anal. **231** , no. 2, pp. 269–286
7. Parisi, G. (1998) *On the probabilistic formulation of the replica approach to spin glasses*, Preprint, Cond-Math/90/1081
8. Baffioni, F., Rosati, F. (2000) *Some exact results on the ultrametric overlap distribution for mean field spin glass models*, EPJ direct B2, 1–17
9. Ghirlanda S., Guerra, F. (1998) *General properties of overlap probability distributions in disordered spin systems. Towards Parisi ultrametricity*, J. Phys. A 31, pp. 9149–9155
10. Talagrand, M. (2006) *The Parisi formula*, Ann. Math. **106**, pp. 221–263
11. Rizzo, T. (2000) *Against chaos in temperature in Mean Field Spin Glass Models*, J. of Phys. A: Mathematical and general, 34, no 27, pp. 5536–5550
12. Rizzo, T., Crisante A. (2003) *Chaos in temperature in the Sherrington–Kirkpatrick model*, Phys. Rev. lett. 90, pp. 137–201
13. Rizzo, T. *Ultrametricity between states at different temperatures in Spin-Glasses*, Cond-Math preprint 0207071

Much Ado about Derrida's GREM

Anton Bovier[1,2] and Irina Kurkova[3]

[1]Weierstraß Institute for Applied Analysis and Stochastics
Mohrenstraße 39, 10117 Berlin, Germany
[2]Institute for Mathematics, Berlin University of Technology
Straße des 17. Juni 136, 10623 Berlin, Germany
e-mail: bovier@wias-berlin.de
[3]Laboratoire de Probabilitiés et Modèles Aléatoires, Université Paris 6
B.C. 188, 4, Place Jussieu, 75252 Paris Cedex 5, France
e-mail: kourkom@ccr.jussieu.fr

Summary. We provide a detailed analysis of Derrida's generalised random energy model (GREM). In particular, we describe its limiting Gibbs measure in terms of Ruelle's Poisson cascades. Next we introduce and analyse a more general class of continuous random energy models (CREMs) which differs from the well-known class of Sherrington–Kirkpatrick models only in the choice of distance on the space of spin configurations: the Hamming distance defines the later class while the ultrametric distance corresponds to the former one. We express explicitly the geometry of its limiting Gibbs measure in terms of genealogies of Neveu's continuous state branching process via an appropriate time change. We also identify the distances between replicas under the limiting CREM's Gibbs measure with those between integers of Bolthausen–Sznitman coalescent under the same time change.

Key words: Gaussian processes, spin-glasses, generalised random energy model, Poisson point processes, branching processes, coalescence.

1 Introduction

Through the remarkable progress achieved recently through the work of Guerra [1], Aizenman, et al. [2], and Talagrand [3, 4] (see also this volume) towards a rigorous justification of the Parisi solution in the Sherrington–Kirkpatrick models, we have now a clear understanding of how Parisi's replica symmetry breaking solution for the free energy emerges. With the exception of a regime in the p-spin SK models where one-step replica symmetry breaking occurs [5, 6], however, these results only justify the formula for the free energy. The question of how the asymptotics of the Gibbs measure is described in general, and whether it conforms to the picture suggested by the replica theory remains open.

2000 *Mathematics Subject Classification.* 82B44.

In this situation it may still be instructive to see how a picture like the one predicted by replica theory emerges in another class of spin glass models, the Generalized Random Energy models of Derrida and Gardner [7–9]. This is reinforced by the fact that these structures play a crucial rôle in the Parisi solution. In this article we give a concise review of a detailed rigorous analysis of the asymptotics of the Gibbs measures in this class of models that we carried out recently [10–12].

The class of models we consider here can be described as follows. Consider the N-dimensional hypercube $\Sigma_N = \{-1, 1\}^N$ endowed with the (normalised) ultrametric distance

$$d_N(\sigma, \tau) = 1 - N^{-1}(\min\{i : \sigma_i \neq \tau_i\} - 1). \tag{1.1}$$

Define a centred normal Gaussian process X on Σ_N with covariance given by

$$\mathbb{E}\, X_\sigma X_\tau = A(1 - d_N(\sigma, \tau)) \tag{1.2}$$

for some non-decreasing right-continuous function $A : [0, 1] \to [0, 1]$.

The principal objects of interest are the Gibbs measures on Σ_N:

$$\mu_{\beta,N}(\sigma) \equiv \frac{e^{\beta\sqrt{N}X_\sigma}}{Z_{\beta,N}}, \qquad \sigma \in \Sigma_N, \tag{1.3}$$

where the partition function, $Z_{\beta,N}$, is

$$Z_{\beta,N} = \sum_{\sigma \in \Sigma_N} e^{\beta\sqrt{N}X_\sigma}. \tag{1.4}$$

This class of models differs from the Sherrington–Kirkpatrick (SK) models *only* in the choice of the distance (1.1). In fact, the SK models are defined in the same way, but instead of the ultrametric distance d_N one uses the Hamming distance,

$$d_N^H(\sigma, \tau) = N^{-1}\#\{i : \sigma_i \neq \tau_i\}.$$

Then the Hamiltonian of the class of SK models has a covariance structure $\mathbb{E}\, X_\sigma X_\tau = A(d_N^H(\sigma, \tau))$ with any function A such that the matrix of $A(d_N^H(\sigma, \tau))$ is positively defined. Since $N^{-1}\sum_i \sigma_i \tau_i = 1 - 2d_N^H(\sigma, \tau)$, the choice of $A(x) = (1 - 2x)^2$ corresponds to the original SK model [13].

1.1 History of the Models

In 1980 Derrida proposed the simplest spin-glass model, where the standard Gaussian random variables X_σ are independent [14, 15]. It was called the *random energy model* (REM). Note that this is a particular case of the model (1.2) with $A(x) = 1_{\{x=1\}}$.

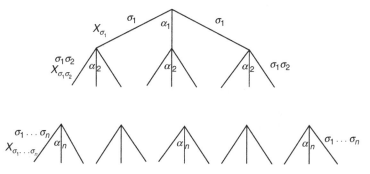

Fig. 1. Structure of the Hamiltonian of the GREM

Derrida also introduced later [7] the *generalised random energy model* (GREM) in view of keeping dependence while simplifying it to a hierarchical structure in order to obtain a more tractable model. The Hamiltonian of the GREM can be constructed explicitly in terms of i.i.d. Gaussian random variables. Namely, chose the parameters $n \geq 1$ (number of hierarchies), $a_1, a_2, \ldots, a_n \in [0,1]$ with $\sum_{i=1}^{n} a_i = 1$, and $\alpha_1, \alpha_2, \ldots, \alpha_n \in [1,2]$ with $\prod_{i=1}^{n} \alpha_i = 2$. Let us represent the hypercube Σ_N as a product $\Sigma_N = \prod_{i=1}^{n} \Sigma_{N \ln \alpha_i / \ln 2}$ and write $\sigma = \sigma_1, \ldots, \sigma_n$ where $\sigma_i \in \Sigma_{N \ln \alpha_i / \ln 2}$. Let $X_{\sigma_1}, X_{\sigma_1 \sigma_2}, \ldots, X_{\sigma_1 \cdots \sigma_n}$ be $\alpha_1^N + \alpha_1^N \alpha_2^N + \cdots + \alpha_1^N \cdots \alpha_n^N$ independent standard Gaussian random variables. Then the Hamiltonian of the GREM is given by:

$$X_\sigma = \sqrt{a_1} X_{\sigma_1} + \sqrt{a_2} X_{\sigma_1 \sigma_2} + \cdots + \sqrt{a_n} X_{\sigma_1 \cdots \sigma_n} \quad \text{if} \quad \sigma = \sigma_1 \ldots \sigma_n. \quad (1.5)$$

To get some intuition in (1.5), one could imagine a tree illustrated on Fig. 1 : α_1^N branches of the first level are indexed by σ_1. Each of these branches supports α_2^N branches of the second level indexed by $\sigma_1 \sigma_2$: thus on the second level there are $(\alpha_1 \alpha_2)^N$ branches, etc. Each configuration $\sigma = \sigma_1 \ldots \sigma_n$ is represented uniquely as a path on this tree going from the top to the bottom through the branches $\sigma_1, \sigma_1 \sigma_2, \ldots, \sigma_1 \ldots \sigma_n$. If, moreover, we associate to each of branches $\sigma_1 \ldots \sigma_k$ a random variable $X_{\sigma_1 \ldots \sigma_k}$, then X_σ is the linear combination of these random variables taken along the path associated with σ and multiplied by coefficients $\sqrt{a_1}, \ldots, \sqrt{a_n}$.

As can be verified by computing the covariance of X_σ, this model is a special case of the models (1.2), where $A(x)$ is a step function given as

$$A(x) = \sum_{i=0}^{k} a_k, \quad \text{for } x \in (\ln(\alpha_0 \cdots \alpha_k)/\ln 2, \ln(\alpha_0 \cdots \alpha_{k+1})/\ln 2), \quad (1.6)$$

$k = 0, 1, \ldots, n$, where $a_0 = 0$, $\alpha_0 = 1$; see Fig. 2. The GREM was analysed by Derrida and Gardner [7–9, 16]. A rigorous computation of the free energy, $N^{-1} \ln \sum_\sigma e^{\beta \sqrt{N} X_\sigma}$, in full generality was later given [17]. Derrida and Gardner [8] also considered limits of their results as the number of steps tended

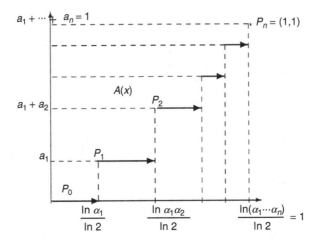

Fig. 2. The function $A(x)$ of the GREM

to infinity, and interpreted these results as corresponding to continuous functions A.

While there were very few further rigorous results on these models (but see [18, 19]), Ruelle in a seminal paper of 1988 [20] introduced a new class of models based on *Poisson cascades* (to which we will henceforth refer to as "Ruelle's REM and GREM"), which he understood to be the appropriate models to describe the limiting Gibbs measures of Derrida's GREMs. Ruelle noted a number of remarkable features of these models, and in particular observed that it was possible to construct limits as the number of steps went to infinity in terms of projective limits.

Shortly after that, Neveu [21] observed a connection between Ruelle's models and *continuous state branching processes*. Unfortunately, this remark appeared only in a preliminary internal report that was never published. Later, Bolthausen and Sznitman in [22] interpreted the results of the replica theory of spin glasses in terms of a coalescent process, now known as the *Bolthausen–Sznitman coalescent*. Following this paper, Bertoin and Le Gall [23] gave a precise and complete form of the relation between Neveu's continuous state branching processes, Ruelle's GREM, and the Bolthausen–Sznitman coalescent.

Around the time when these fascinating results appeared, we began to investigate more closely the link to the original spin-glass models with Ruelle's models. In the REM, this connection was made in a paper with Löwe [24] (see also [25] and [6, 26, 27]). These results were extended to the GREMs in [10], using essentially elementary methods. We observed, however, that the use of the so-called Ghirlanda–Guerra identities [28] allowed for a different approach that circumvents parts of these explicit computations (this fact was first observed in the REM by Talagrand [27], who also exploited these identities heavily in his work on the p-spin SK models [5,6,27,29]). It allowed us in [11] to extend our convergence results to the general class of models defined above with general right-continuous non-decreasing functions $A(x) : [0,1] \to [0,1]$. We called this class *continuous random energy models* (CREM). Finally, combining the results of [11] and those of Bertoin and Le Gall [23], we concluded our investigation in [12] by linking our results to the continuous state branching process of Neveu. More precisely, we identified the geometry of the limiting Gibbs measure proven to exist in [11] explicitly in terms of the genealogy of Neveu's branching process, which were defined in [23]. The role played by these random genealogies in the Parisi solution can be most clearly seen in the paper by Aizenman, Sims, and Starr [2] (see also this volume). We hope that these examples help to explain to a mathematical audience what physicist describe when they talk about "continuous replica symmetry breaking".

1.2 Geometry of Gibbs Measures

The central problem one is faced with when analysing mean field spin glasses is to describe the geometric structure of a random probability measure (1.3) on a set Σ_N as $N \to \infty$. Two scenarios can be expected

1. At high temperatures (small β) this measure will be spread over an exponentially large set of configurations that is distributed rather uniformly over the hypercube.
2. At low temperatures (large β) this measure will concentrate on a very small subset of configurations σ, with a rather complicated structure, corresponding to the largest values of X_σ, while the mass of the enormous amount of all other configurations σ will be negligible.

These statements are easily proven in the REM, using the classical theory of extremes of i.i.d. random variables [30]. Let us briefly recall these results. To be able to embed all hypercubes Σ_N, $N \in \mathbb{N}$, in the same compact space, it is convenient to map them to the unit interval via the canonical maps $r_N : \Sigma_N \to [0,1]$:

$$r_N(\sigma) = 1 - \sum_{i=1}^{N} 2^{-i-1}(1 + \sigma_i). \tag{1.7}$$

For finite N, the Gibbs measure is then mapped to a discrete measure on $[0,1]$ concentrated on 2^N points:

$$\widetilde{\mu}_{\beta,N} = \sum_{\sigma \in \Sigma_N} \mu_{\beta,N}(\sigma)\delta_{r_N(\sigma)} \tag{1.8}$$

with distribution function

$$\theta_{\beta,N}(x) = \int_0^x \mathrm{d}\,\widetilde{\mu}_{\beta,N}. \tag{1.9}$$

It was proved in [25] that

$$
\theta_{\beta,N} \xrightarrow{\mathcal{D}}
\begin{cases}
y = x & \text{if } \beta \leq \sqrt{2\ln 2} \\
\dfrac{S_{\beta/\sqrt{2\ln 2}}(x)}{S_{\beta/\sqrt{2\ln 2}}(1)} & \text{if } \beta > \sqrt{2\ln 2}.
\end{cases}
\tag{1.10}
$$

This means that $\widetilde{\mu}_{\beta,N}$ converges to the Lebesgue measure on $[0,1]$ at high temperatures, confirming scenario (1). The random function $S_{\beta/\sqrt{2\ln 2}}(x)$ is a stable subordinator with the index $\beta/\sqrt{2\ln 2}$, i.e. a step function that jumps at random points, t_i, $i = 1, 2, \ldots$, which are distributed uniformly on $[0,1]$. The values of jumps w_i are also random and can be expressed as

$$
w_i = \frac{e^{(\beta/\sqrt{2\ln 2})x_i}}{\sum_j e^{(\beta/\sqrt{2\ln 2})x_j}},
\tag{1.11}
$$

where $x_1 > x_2 > \cdots$ are the atoms of the Poisson point process $\mathcal{P}$ on $\mathbb{R}$ with intensity measure $e^{-x}dx$. This confirms scenario (2): at low temperatures the limiting Gibbs measure concentrates on a countable number of randomly chosen configurations corresponding to points $t_i \in [0,1]$.

This description of the limiting Gibbs measure does not give any information about its geometry. But to define the geometry of a measure on the infinite dimensional hypercube, it is necessary, first of all, to specify a topology. The conventional choice of the product topology is not suitable to capture the fact that these measures tend to concentrate on *individual random* configurations. To resolve this problem we introduce the following construction. Let

$$
m_\sigma(t) = \mu_{\beta,N}(\tau : d_N(\sigma,\tau) \leq 1 - t)
\tag{1.12}
$$

be the picture of the landscape of the Gibbs measure taken from a given configuration σ. The function $1 - m_\sigma(t)$ is a random distribution function on $[0,1]$. In this way we get 2^N different pictures of the landscape of the Gibbs measure taken from different configurations σ. It seems reasonable to subject the importance of each of these pictures to the Gibbs mass, $\mu_{\beta,N}(\sigma)$, of its starting point, σ. In this way we construct the random probability measure

$$
\mathcal{K}_{\beta,N} \equiv \sum_{\sigma \in \Sigma_N} \mu_{\beta,N}(\sigma)\delta_{m_\sigma}(\cdot)
\tag{1.13}
$$

on these distribution functions $m_\sigma(t)$ that attributes to each function $m_\sigma(t)$ the weight $\mu_{\beta,N}(\sigma)$. We call $\mathcal{K}_{\beta,N}$ the *empirical distance distribution function*. It has a very appealing physical interpretation: it tells, for a fixed realisation of the disorder, with which probability an observer, that is himself distributed with the Gibbs measure, will see a given distribution of mass around himself. Convergence results for the Gibbs measures will be formulated in term of convergence of the law, under the Gaussian process X_σ, of $\mathcal{K}_{\beta,N}$. A key object is the first moment of $\mathcal{K}_{\beta,N}$:

$$
\int \mathcal{K}_{\beta,N}(dm)m(\cdot) = \mu_{\beta,N}^{\otimes 2}(\sigma,\tau : d_N(\sigma,\tau) \in \cdot),
\tag{1.14}
$$

which is the probability that two configurations, σ, τ, drawn independently from the Gibbs sample satisfy $d_N(\sigma, \sigma') \in$.

In the case of the REM, the limit of $\mathcal{K}_{\beta,N}$ is a rather simple object:

$$
\mathcal{K}_{\beta,N} \xrightarrow{\mathcal{D}}
\begin{cases}
\delta_{\delta(0)} & \beta \le \sqrt{2\ln 2} \\
\sum_{w_i} w_i \delta_{w_i \delta(0) + (1-w_i)\delta(1)} & \beta > \sqrt{2\ln 2}.
\end{cases}
\tag{1.15}
$$

It will manifest much more rich and interesting structure in the case of the GREM and CREM, as we will see.

1.3 Point Process of Extremes

To describe efficiently the behaviour of the limiting Gibbs measure according to scenario (2), it is necessary to know the maximal values of the Gaussian process X_σ. In the case of *independent* variables the corresponding result is well known. First of all $\max_{\sigma \in \Sigma_N} X_\sigma N^{-1/2} \to \sqrt{2\ln 2}$ a.s. from where for any $\epsilon > 0$

$$
\mathsf{P}(\forall \sigma : \ X_\sigma < \sqrt{N}(\sqrt{2\ln 2} + \epsilon)) \to 1, \quad \mathsf{P}(\forall \sigma : \ X_\sigma < \sqrt{N}(\sqrt{2\ln 2} - \epsilon)) \to 0.
$$

To get the limiting value here between 0 and 1, one should take $\epsilon = \epsilon_N \to 0$ as $N \to \infty$. It turns out that the right function depending on the parameter $x \in \mathbb{R}$ is

$$
u_{\ln \alpha, N}(x) = \sqrt{2N\ln \alpha} + \frac{x}{\sqrt{2N\ln \alpha}} - \frac{\ln N + \ln \ln \alpha + \ln 4\pi}{2\sqrt{2N\ln \alpha}},
\tag{1.16}
$$

with $\alpha = 2$ as

$$
\mathsf{P}(\forall \sigma : \ X_\sigma < u_{\ln 2, N}(x)) \to e^{-e^{-x}}, \quad N \to \infty.
$$

Thus, we come to the classical result on the convergence of extreme value statistics in the case where X_σ are 2^N independent Gaussian random variables. It says that the point process

$$
\sum_{\sigma \in \Sigma_N} \delta_{u_{\ln 2, N}^{-1}(X_\sigma)} \xrightarrow{\mathcal{D}} \mathcal{P}
\tag{1.17}
$$

converges weakly to the Poisson point process $\mathcal{P}$ on $\mathbb{R}$ with the intensity measure $e^{-x}dx$, see e.g. [30]. This result is the crucial ingredient in the proof of (1.10) and clarifies the meaning of (1.11).

To start the analysis of the GREM, we need an analogous result in the case of correlated Gaussian random variables. Results of this kind in the correlated case are much more scarce. Most of them establish conditions under which the same limiting point process arises as for the independent random variables. We will see that this is in general not the case for the random variables (1.5) correlated as in the Hamiltonian of the GREM.

1.4 Organisation of the Paper

The remainder of the paper is organised as follows. Sect. 2 is devoted to convergent point processes associated with the Hamiltonian of the GREM. Namely, we find the point process of extreme value statistics of its Hamiltonian. These results can be viewed as those on convergence of extreme value statistics for correlated Gaussian random variables independently of the context of spin glasses. In Sect. 3 we study the GREM (1.5) with finitely many hierarchies. In particular we identify the limit of $\mathcal{K}_{\beta,N}$ for this model with Ruelle's probability cascades. In Sect. 4 we analyse the general case of CREM's (1.1), (1.2) with a "continuum of hierarchies". We prove the existence of the limit of $\mathcal{K}_{\beta,N}$ by the so-called "Ghirlanda–Guerra" identities, i.e. identifying limits of all its moments. In Sect. 5 we describe explicitly the limit of $\mathcal{K}_{\beta,N}$ in terms of the genealogical structure of Neveu's continuous state branching process modulo an appropriate time change depending only on β and on the concave hull of A.

Notations. When A is a step-function as on Fig. 2, we will denote by $\underline{A}(x)$ its linear interpolation. Its graph consists of the segments $[P_0, P_1]$, $[P_1, P_2]$, ..., $[P_{n-1}, P_n]$ where $P_k = (\sum_{i=0}^{k} a_k, \ln(\alpha_0 \cdots \alpha_k)/\ln 2)$ for $k = 0, \ldots, n$, with $a_0 = 0$, $\alpha_0 = 1$ so that $P_0 = (0,0)$ and $P_n = (1,1)$, see Fig. 2.

We will denote by $\widehat{A}(x)$ the concave hull of the function $A(x)$ and by $\widehat{A}'(x)$ the right derivative of the concave hull of A, see Fig. 5.

2 Convergent Point Processes Associated to the GREM

In Theorem 2.1 we give a necessary and sufficient condition on the parameters a_i, α_i which assure that point process of extreme value statistics of GREM's Hamiltonian (1.5) is the same as in the case of independent random variables (1.17). This condition is the convexity of the linear interpolation $\underline{A}(x)$. In other words, the concave hull of $\underline{A}(x)$ should be the straight line $y = x$. It is illustrated on Fig. 3a. This condition is strictly weaker than the sufficient condition implied by Slepian's lemma on the comparison of Gaussians. (Theorem 4.2.1 in [30]). We use the notation (1.16).

Theorem 2.1. [10] *Let* $n \in \mathbb{N}$, $n \geq 1$, $0 < a_i < 1$ *with* $\sum_{i=1}^{n} a_i = 1$, $\alpha_i > 1$, $i = 1, 2, \ldots, n$. *The point process*

$$\sum_{\sigma = \Sigma_N} \delta_{u_{\ln 2, N}^{-1}\left(\sqrt{a_1} X_{\sigma_1} + \sqrt{a_2} X_{\sigma_1 \sigma_2} + \cdots + \sqrt{a_n} X_{\sigma_1 \sigma_2 \cdots \sigma_n}\right)}$$

converges weakly to the Poisson point process $\mathcal{P}$ *on* $\mathbb{R}$ *with the intensity measure* $K e^{-x} dx$, $K \in \mathbb{R}$, *if the linear interpolation* $\underline{A}(x)$ *is convex, that is*

$$a_i + a_{i+1} + \cdots + a_n \geq \ln(\alpha_i \alpha_{i+1} \cdots \alpha_n)/\ln \bar{\alpha} \quad \text{for all} \quad i = 2, 3, \ldots, n, \quad (2.1)$$

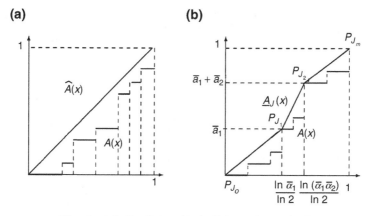

Fig. 3. (a) Condition (2.1), **(b)** condition (2.5)

see Fig. 3a. If all inequalities in (2.1) are strict, then $K = 1$. If some of the relations are equalities, then

$$0 < K < 1.^1$$

The next lemma gives a sufficient condition for the convergence of the multi-dimensional point process to the point process of Poisson cascades defined by Ruelle in [20]. This is a generalisation of Theorem 3 of [19]: we do not specify the law of the vectors $Y_{\sigma_1\ldots\sigma_i}$, neither assume their independence.

Lemma 2.1. [10] Let $\alpha_i \geq 1$, $i = 1, 2, \ldots, k$, $\bar{\alpha} \equiv \prod_{i=1}^{k} \alpha_i$. Let Y_{σ_1}, $Y_{\sigma_1\sigma_2}, \ldots, Y_{\sigma_1\ldots\sigma_k}$ be $\alpha_1^N + \cdots + (\alpha_1 \cdots \alpha_k)^N$ identically distributed random variables. Assume that $1 + \alpha_1^N + \cdots + (\alpha_1 \cdots \alpha_{k-1})^N$ vectors $(Y_{\sigma_1})_{\sigma_1 \in \{-1,1\}^{N \ln \alpha_1 / \ln \bar{\alpha}}}, (Y_{\sigma_1\sigma_2})_{\sigma_2 \in \{-1,1\}^{N \ln \alpha_2 / \ln \bar{\alpha}}} \forall \sigma_1 \in \{-1,1\}^{N \ln \alpha_1 / \ln \bar{\alpha}}$, $\ldots, (Y_{\sigma_1\sigma_2\ldots\sigma_k})_{\sigma_k \in \{-1,1\}^{N \ln \alpha_k / \ln \bar{\alpha}}} \forall \sigma_1 \ldots \sigma_{k-1} \in \{-1,1\}^{N \ln(\alpha_1 \cdots \alpha_{k-1}) / \ln \bar{\alpha}}$ are independent.

Let $v_{N,1}(x), \ldots, v_{N,k}(x)$ be functions on $\mathbb{R}$ such that the following point processes

$$\sum_{\sigma_1} \delta_{v_{N,1}(Y_{\sigma_1})} \xrightarrow{\mathcal{D}} \mathcal{P}_1$$

$$\sum_{\sigma_2} \delta_{v_{N,2}(Y_{\sigma_1\sigma_2})} \xrightarrow{\mathcal{D}} \mathcal{P}_2 \quad \forall \sigma_1$$

$$\cdots$$

$$\sum_{\sigma_k} \delta_{v_{N,k}(Y_{\sigma_1\sigma_2\ldots\sigma_k})} \xrightarrow{\mathcal{D}} \mathcal{P}_k \quad \forall \sigma_1 \ldots \sigma_{k-1}$$

(2.2)

---

1 Explicit expressions for K are given in [10].

converge weakly to the Poisson point processes $\mathcal{P}_1, \ldots, \mathcal{P}_k$ *on* $\mathbb{R}$ *with the intensity measures* $K_1 e^{-x} dx, \ldots, K_k e^{-x} dx$ *with some constants* $K_1, \ldots, K_k > 0$, *respectively. Then the following point process on* $\mathbb{R}^k$:

$$\mathcal{P}_N^{(k)} \equiv \sum_{\sigma_1} \delta_{v_{N,1}(Y_{\sigma_1})} \sum_{\sigma_2} \delta_{v_{N,2}(Y_{\sigma_1 \sigma_2})} \cdots \sum_{\sigma_k} \delta_{v_{N,k}(Y_{\sigma_1 \sigma_2 \cdots \sigma_k})} \xrightarrow{\mathcal{D}} \mathcal{P}^{(k)}$$

converges weakly to a point process, $\mathcal{P}^{(k)}$, *called a* k-level Poisson cascade, *on* $\mathbb{R}^k$.

Structure of $\mathcal{P}^{(k)}$. The Poisson cascades $\mathcal{P}^{(k)}$ can be characterised in terms of their Laplace transforms, see [10]. Informally, they are best described as follows [20]: If $k = 1$, it is a ordinary Poisson point process on $\mathbb{R}$ with intensity measure $K_1 e^{-x} dx$. To construct $\mathcal{P}^2$ on $\mathbb{R}^2$, we place the process $\mathcal{P}^1$ for $k = 1$ on the axis of the first coordinate and through each of its points draw a straight line parallel to the axis of the second coordinate. Then we put on each of these lines independently a Poisson point process with intensity $K_2 e^{-x} dx$. These points on $\mathbb{R}^2$ form the process $\mathcal{P}^2$. This procedure is now simply iterated k times.

Theorem 2.1 and Lemma 2.1 combined to give a first important result, that establishes which convergent point processes may be constructed in the GREM: one can group together the hierarchies between the levels $J_0, J_1, \ldots, J_m$, if condition (2.5) is verified. This condition is illustrated in Fig. 3b: it means the convexity of the function $\underline{A}(x)$ between the levels $J_0, J_1, \ldots, J_m$.

Theorem 2.2. [10] *Let* $\alpha_i \geq 1$, $0 < a_i < 1$, $i = 1, 2, \ldots, n$, $\prod_{i=1}^n \alpha_i = 2$, $\sum_{i=1}^n a_i = 1$. *Let* $J_1, J_2, \ldots, J_m \in \mathbb{N}$ *be the indices* $0 = J_0 < J_1 < J_2 < \cdots < J_m = n$. *We denote by* $\bar{a}_l \equiv \sum_{i=J_{l-1}+1}^{J_l} a_i$, $\bar{\alpha}_l \equiv \prod_{i=J_{l-1}+1}^{J_l} \alpha_i$, $l = 1, 2, \ldots, m$, *and introduce the standard Gaussian random variables*

$$\bar{X}^{\sigma_1 \ldots \sigma_{J_{l-1}}}_{\sigma_{J_{l-1}+1} \sigma_{J_{l-1}+2} \cdots \sigma_{J_l}} \equiv \left(\sqrt{a_{J_{l-1}+1}} X_{\sigma_1 \ldots \sigma_{J_{l-1}} \sigma_{J_{l-1}+1}} \right.$$
$$+ \sqrt{a_{J_{l-1}+2}} X_{\sigma_1 \ldots \sigma_{J_{l-1}} \sigma_{J_{l-1}+1} \sigma_{J_{l-1}+2}}$$
$$\left. + \cdots + \sqrt{a_{J_l}} X_{\sigma_1 \ldots \sigma_{J_{l-1}} \sigma_{J_{l-1}+1} \ldots \sigma_{J_l}} \right) / \sqrt{\bar{a}_l}. \qquad (2.3)$$

Assume that a partition $J_1, J_2, \ldots, J_m$ *satisfies the following condition: for all* $l = 1, 2, \ldots, m$ *and all* k *such that* $J_{l-1} + 2 \leq k \leq J_l$

$$(a_k + a_{k+1} \cdots + a_{J_l-1} + a_{J_l}) / \bar{a}_l \geq \ln(\alpha_k \alpha_{k+1} \cdots \alpha_{J_l-1} \alpha_{J_l}) / \ln(\bar{\alpha}_l). \qquad (2.4)$$

If $\underline{A}^J(x)$ *is the linear interpolation of the points* $(0,0), P_{J_1}, P_{J_2}, \ldots, P_{J_m} = (1,1)$, *condition (2.4) is equivalent to*

$$\underline{A}(x) \leq \underline{A}^J(x) \quad \forall x \in [0,1], \qquad (2.5)$$

(see Fig. 3b), then the point process

$$\mathcal{P}_N^{(m)} \equiv \sum_{\sigma_1 \cdots \sigma_{J_1}} \delta_{u_{\ln \bar\alpha_1, N}^{-1}(\bar X_{\sigma_1 \cdots \sigma_{J_1}})} \sum_{\sigma_{J_1+1} \cdots \sigma_{J_2}} \delta_{u_{\ln \bar\alpha_2, N}^{-1}(\bar X_{\sigma_{J_1+1} \cdots \sigma_{J_2}}^{\sigma_1 \cdots \sigma_{J_1}})} \cdots \sum_{\sigma_{J_{m-1}+1} \cdots \sigma_{J_m}}$$

$$\delta_{u_{\ln \bar\alpha_m, N}^{-1}(\bar X_{\sigma_{J_{m-1}+1} \cdots \sigma_{J_m}}^{\sigma_1 \cdots \sigma_{J_{m-1}}})} \tag{2.6}$$

converges weakly in distribution to the point process $\mathcal{P}^{(m)}$ on $\mathbb{R}^m$, defined in Lemma 2.1, with constants $K_1, \ldots, K_m$. Moreover, $K_l = 1$, if all $J_l - J_{l-1} - 1$ inequalities in (2.4) for $k = J_{l-1}+2, \ldots, J_l$ are strict. Otherwise, $0 < K_l < 1$.[2]

It is clear that the point process of extreme values of the Hamiltonian can be constructed from one of the partitions of Theorem 2.2. This is the one that allows to group together the maximal number of hierarchies: among all series of indices $J_1, \ldots, J_m$ satisfying (2.4) one should choose the one with the largest differences $J_1 - J_0, \ldots, J_m - J_{m-1}$. To define it, we set $J_0 \equiv 0$ and

$$J_l \equiv \min\{J > J_{l-1} : A_{J_{l-1}+1, J} > A_{J+1, k} \ \forall k \geq J + 1\}$$

$$where \, A_{j,k} \equiv \frac{\sum_{i=j}^k a_i}{2 \ln(\prod_{i=j}^k \alpha_i)}. \tag{2.7}$$

The sequence $J_1, \ldots, J_m$, defined by (2.7), verifies (2.4), for all k, such that $J_{l-1} + 2 \leq k \leq J_l$ and all $l = 1, 2, \ldots, m$. This choice of the partition $J_1, J_2, \ldots, J_m$ (2.7) has a beautiful geometric interpretation: the linear interpolation $\underline{A}^J(x)$ of $(0,0), P_{J_1}, \ldots, P_{J_m} = (1,1)$ is *the concave hull* of the function $\underline{A}(x)$, see Fig. 4. We set

$$\bar a_l \equiv \sum_{i=J_{l-1}+1}^{J_l} a_i, \quad \bar\alpha_l \equiv \prod_{i=J_{l-1}+1}^{J_l} \alpha_i, \quad \gamma_l \equiv \sqrt{\frac{\bar a_l}{2 \ln \bar\alpha_l}} = \sqrt{\frac{(\widehat A)'(P_{J_{l-1}})}{2 \ln 2}},$$

$$l = 1, 2, \ldots, m, \tag{2.8}$$

see Fig. 4. Next, let us define the function $U_{J,N}$ as

$$U_{J,N}(x) \equiv \sum_{l=1}^m \left(\sqrt{2 N \bar a_l \ln \bar\alpha_l} - N^{-1/2} \gamma_l (\ln(N(\ln \bar\alpha_l)) + \ln 4\pi)/2 \right) + N^{-1/2} x \tag{2.9}$$

and the point process

$$\mathcal{E}_N \equiv \sum_{\sigma \in \{-1,1\}^N} \delta_{U_{J,N}^{-1}(\sqrt{a_1} X_{\sigma_1} + \cdots + \sqrt{a_n} X_{\sigma_1 \cdots \sigma_n})}. \tag{2.10}$$

---

[2]Explicit expressions for K are given in [10].

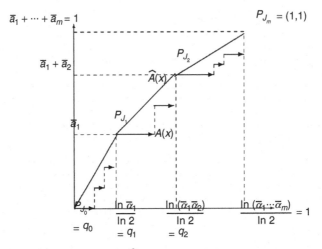

Fig. 4. The function $\widehat{A}(x)$, with parameters (2.8), (3.6).

Theorem 2.3. [10] *(i) The point process $\mathcal{E}_N$ converges weakly, as $N \uparrow \infty$, to the point process on $\mathbb{R}$*

$$\mathcal{E} \equiv \int_{\mathbb{R}^m} \mathcal{P}^{(m)}(dx_1, \ldots, dx_m) \delta_{\sum_{l=1}^{m} \gamma_l x_l} \tag{2.11}$$

where $\mathcal{P}^{(m)}$ is a Poisson cascade (introduced in Lemma 2.1) with constants $K_1, \ldots, K_m$, as defined in Theorem 2.2 according to the partition $J_1, \ldots, J_m$ of (2.7) and the parameters $\gamma_1, \ldots, \gamma_m$ defined by (2.8).
(ii) The inequalities $\gamma_1 > \cdots > \gamma_m$ imply the existence of $\mathcal{E}$.
(iii) We have $\max_\sigma(X_\sigma/\sqrt{N}) \to \sqrt{\bar{a}_1 2 \ln \bar{\alpha}_1} + \cdots + \sqrt{\bar{a}_m 2 \ln \bar{\alpha}_m}$ a.s. and also $\mathsf{E}(\max_\sigma X_\sigma/\sqrt{N}) \to \sqrt{2\bar{a}_1 \ln \bar{\alpha}_1} + \cdots + \sqrt{2\bar{a}_m \ln \bar{\alpha}_m}$.

3 GREM: Detailed Analysis

3.1 Fluctuations of the Partition Function

For any sequence of indices $0 < J_1 < \cdots < J_m = n$, the partition function (1.4) of the GREM can be written as:

$$
\begin{aligned}
Z_{\beta,N} = e^{\sum_{j=1}^{m} \left(\beta N \sqrt{2\bar{a}_j \ln \bar{\alpha}_j} - \beta \gamma_j [\ln(N \ln \bar{\alpha}_j) + \ln 4\pi]/2\right)} \\
\times \sum_{\sigma_1 \cdots \sigma_{J_1}} e^{\beta \gamma_1 u_{\ln \bar{\alpha}_1, N}^{-1}(\bar{X}_{\sigma_1 \cdots \sigma_{J_1}})} \cdots \\
\sum_{\sigma_{J_{m-1}+1} \cdots \sigma_{J_m}} e^{\beta \gamma_m u_{\ln \bar{\alpha}_m, N}^{-1}(\bar{X}_{\sigma_{J_{m-1}+1} \cdots \sigma_{J_m}}^{\sigma_1 \cdots \sigma_{J_{m-1}}})},
\end{aligned} \tag{3.1}
$$

where $\bar{a}_l \equiv \sum_{i=J_{l-1}+1}^{J_l} a_i$, $\bar{\alpha}_l \equiv \prod_{i=J_{l-1}+1}^{J_l} \alpha_i$, $\gamma_l \equiv \sqrt{\bar{a}_l}/\sqrt{2\ln \bar{\alpha}_l}$, $l = 1, 2, \ldots, m$, and the random variables $\bar{X}_{\sigma_{J_{l-1}+1} \ldots \sigma_{J_l}}^{\sigma_1 \ldots \sigma_{J_{l-1}}}$ are defined in (2.3). For any sequence $J_1, \ldots, J_m$, satisfying conditions (2.4), the point process (2.6) in the exponent of (3.1) converges to the corresponding Poisson cascade by Theorem 2.2. The sequence constructed according to (2.7) gives the correct scale of fluctuations of $Z_{\beta,N}$ via (3.1). Nevertheless it should be cut at a certain level $J_{l(\beta)}$ that depends on the temperature: using the sequence $\gamma_1 > \gamma_2 > \cdots > \gamma_m$ defined in (2.8), we set

$$l(\beta) \equiv \max\{l \geq 1 : \beta\gamma_l > 1\} \tag{3.2}$$

and $l(\beta) \equiv 0$ if $\beta\gamma_1 \leq 1$. This definition (3.2) has a simple geometric interpretation:

$$l(\beta) \equiv \max\left\{l \geq 1 : \beta\sqrt{\frac{(\hat{A})'(P_{J_{l-1}})}{2\ln 2}} > 1\right\}.$$

In [17], the limit of the free energy has been computed in terms of (2.8) and (3.2):

$$\lim_{N\to\infty} N^{-1} \ln Z_{N,\beta} = \beta\left(\sqrt{2\bar{a}_1 \ln \bar{\alpha}_1} + \cdots + \sqrt{2\bar{a}_{l(\beta)} \ln \bar{\alpha}_{l(\beta)}}\right)$$

$$+ \sum_{i=J_{l(\beta)}+1}^{n} (\beta^2 a_i/2 + \ln\alpha_j), \text{ a.s.} \tag{3.3}$$

We see that the domain $\{\beta : l(\beta) = 0\} = \{\beta : \beta \leq 1/\gamma_1\}$ is the high temperature region, where $\lim_{N\to\infty} = \frac{1}{N}\mathsf{E}\ln Z_{\beta,N} = \lim_{N\to\infty} \frac{1}{N}\ln \mathsf{E} Z_{\beta,N}$. Theorem 3.1 gives the fluctuations of the partition function.

Theorem 3.1. [10] Let $\alpha_i \geq 1$, $0 < a_i < 1$, $i = 1, 2, \ldots, n$, $\prod_{i=1}^{n} \alpha_i = 2$, $\sum_{i=1}^{n} a_i = 1$. Let $J_1, J_2, \ldots, J_m \in \mathbb{N}$ be the sequence of indices defined by (2.7), the parameters $\bar{a}_i, \bar{\alpha}_i, \gamma_i$ be defined by (2.8) and $l(\beta)$ be defined by (3.2). If $l(\beta) = 0$, then $\frac{Z_{\beta,N}}{2^N e^{\beta^2 N/2}} \to C(\beta)$. If $l(\beta) > 1$, then

$$e^{\sum_{j=1}^{l(\beta)}\left(-\beta N\sqrt{2\bar{a}_j \ln \bar{\alpha}_j} + \beta\gamma_j [\ln(N\ln\bar{\alpha}_j) + \ln 4\pi]/2\right) - N\sum_{i=J_{l(\beta)}+1}^{n}(\beta^2 a_i/2 + \ln\alpha_j)} Z_{\beta,N}$$

$$\xrightarrow{\mathcal{D}} C(\beta)\int_{\mathbb{R}^{l(\beta)}} e^{\beta\gamma_1 x_1 + \beta\gamma_2 x_2 + \cdots + \beta\gamma_{l(\beta)} x_{l(\beta)}} \mathcal{P}^{(l(\beta))}(dx_1 \ldots dx_{l(\beta)}). \tag{3.4}$$

This integral is computed over the Poisson cascades $\mathcal{P}^{(l(\beta))}$ on $\mathbb{R}^{l(\beta)}$, defined in Lemma 2.1, with the constants K_j of Theorem 2.2. The constant

$$C(\beta) = 1, \quad \text{if } \beta\gamma_{l(\beta)+1} < 1, \tag{3.5}$$

and $0 < C(\beta) < 1$, if $\beta\gamma_{l(\beta)+1} = 1$.[3]

---

[3] Explicit formulae for $C(\beta)$ are given in [10]

3.2 Gibbs Measure: Approach via Ruelle's Probability Cascades

We consider everywhere below $\bar{a}_i$, $\bar{\alpha}_i$, γ_i defined by (2.8) according to (2.7) and $l(\beta)$ defined by (3.2). Let us denote the jump points of the derivative of the concave hull $\widehat{A}'(x)$ by

$$q_l \equiv \sum_{n=1}^{l} \frac{\ln \bar{\alpha}_n}{\ln 2}, \quad l = 1, 2, \ldots, m, \tag{3.6}$$

with the convention $q_0 = 0$. They are illustrated on Fig. 4. Let $B_l(\sigma)$ be the ball in Σ_N with centre σ and radius $1 - q$:

$$B_l(\sigma) \equiv \{\sigma' \in \Sigma_N : d_N(\sigma, \sigma') \leq 1 - q_l\} = \{\sigma' : \sigma'_1 \ldots \sigma'_{J_l} = \sigma_1 \ldots \sigma_{J_l}\},$$
$$l = 1, 2, \ldots, l(\beta). \tag{3.7}$$

Let us define the point process $\mathcal{W}_{\beta,N}^{(m)}$ on $(0, 1]^m$ as

$$\mathcal{W}_{\beta,N}^{(m)} \equiv \sum_{\sigma} \delta_{\left(\mu_{\beta,N}(B_1(\sigma)), \ldots, \mu_{\beta,N}(B_m(\sigma))\right)} \frac{\mu_{\beta,N}(\sigma)}{\mu_{\beta,N}(B_m(\sigma))} \tag{3.8}$$

and its projection on the last coordinate

$$\mathcal{R}_{\beta,N}^{(m)} \equiv \sum_{\sigma} \delta_{\mu_{\beta,N}(B_m(\sigma))} \frac{\mu_{\beta,N}(\sigma)}{\mu_{\beta,N}(B_m(\sigma))}. \tag{3.9}$$

It is easy to see that $\mathcal{W}_{\beta,N}^{(m)}$ satisfy the following relation:

$$\mathcal{W}_{\beta,N}^{(m)}(dw_1, \ldots, dw_m) = \int_0^1 \mathcal{W}_{\beta,N}^{(m+1)}(dw_1, \ldots, dw_m, dw_{m+1}) \frac{w_{m+1}}{w_m},$$

where the integral is taken over the last coordinate w_{m+1}. Theorem 3.2 gives the limits of these point processes for all $m \leq l(\beta)$.

Theorem 3.2. [10] *Let $l(\beta) \geq 1$ i.e. $\beta > \sqrt{2 \ln 2 / (\widehat{A})'(0)}$. If $m \leq l(\beta)$, then the point process $\mathcal{W}_{\beta,N}^{(m)}$ on $(0, 1]^m$ converges weakly, as $N \to \infty$, to the point process $\mathcal{W}_{\beta\overline{\gamma}}^{(m)}$, whose atoms $w(i)$ are expressed through the points $(x_1(i), \ldots, x_m(i))$ of the Poisson cascade $\mathcal{P}^{(m)}$ of Lemma 2.1, with constants K_j of Theorem 2.2, as follows:*

$$(w_1(i), \ldots, w_m(i))$$
$$= \left(\frac{\int \mathcal{P}^{(m)}(dy) \delta(y_1 - x_1(i)) e^{\beta(\gamma, y)}}{\int \mathcal{P}^{(m)}(dy) e^{\beta(\gamma, y)}}, \ldots, \frac{\int \mathcal{P}^{(m)}(dy) \delta(y_1 - x_1(i)) \ldots \delta(y_m - x_m(i)) e^{\beta(\gamma, y)}}{\int \mathcal{P}^{(m)}(dy) e^{\beta(\gamma, y)}} \right).$$
$$\tag{3.10}$$

The vector $\overline{\gamma} = (\gamma_1, \ldots, \gamma_m)$ is defined by (2.8) according to (2.7). The process $\mathcal{R}_{\beta,N}^{(m)}$ converges to the process $\mathcal{R}_{\beta}^{(m)}$, where the atoms are the last components of the atoms of (3.10).

The balls $B_{l(\beta)}(\sigma)$ are the smallest ones that have positive mass, $\mu_{\beta,N}$, as $N \to \infty$: For $m > l(\beta)$, $\mu_{\beta,N}(B_m(\sigma)) \to 0$ for any $\sigma \in \Sigma_N$. If $J_{l(\beta)} = n$, i.e. $\beta > 1/\gamma_m = \sqrt{2\ln 2/\lim_{x\to 1}(\widehat{A})'(x)}$, these balls consist of a single configuration, σ. In this case the mass of the Gibbs measure is concentrated on certain randomly chosen individual configurations. Otherwise, these balls consist of all configurations having the same spins as σ starting from the first sit up to the $J_{l(\beta)}$th site.

Definition 3.1. [20] *The process* $\mathcal{W}^{(m)}_{\beta\overrightarrow{\gamma}}$ *defined in Theorem 3.2 is called the process of probability cascades on* $[0,1]^m$ *with m levels and parameters* $\beta\gamma_1 > \cdots > \beta\gamma_m > 1$. I

The most complete object of Theorem 3.2 is of course the process $\mathcal{W}^{(l(\beta))}_{\beta,N}$. Thus, Theorem 3.2 asserts the convergence of the point process $\mathcal{W}^{(l(\beta))}_{\beta,N}$ of Derrida's model with parameters $n \geq 1$, a_i, α_i to the point process of probability cascades of Ruelle's model with parameters $l(\beta)$ and $\beta\gamma_1, \ldots, \beta\gamma_{l(\beta)}$ defined by (2.8) and (3.2). Let us also emphasize the fact that the parameters of the limiting process of probability cascades depend only on the concave hull $\widehat{A}(x)$ and on β.

3.3 Distribution of the Overlaps

One of the most important physical objects is the distribution of the overlap

$$\frac{\sigma \cdot \sigma'}{N} = \frac{\sum_{i=1}^N \sigma_i \sigma'_i}{N} \tag{3.11}$$

of two spin configurations under the Gibbs measure:

$$\widetilde{f}_{\beta,N}(q) \equiv \mu^{\otimes 2}_{\beta,N}\left(\frac{(\sigma \cdot \sigma')}{N} \leq q\right). \tag{3.12}$$

In the context of the GREM it appears more natural to consider the ultrametric distance

$$f_{\beta,N}(q) \equiv \mu^{\otimes 2}_{\beta,N}\left(d_N(\sigma,\sigma') \geq 1 - q\right). \tag{3.13}$$

Theorem 3.3 asserts the remarkable fact that the laws of these two objects coincides in the thermodynamic limit.

Theorem 3.3. [10] *The distribution functions* $f_{\beta,N}$ *et* $\widetilde{f}_{\beta,N}$ *converge in law to the same distribution function* f_β *as* $N \to \infty$. *Moreover* $\mathsf{E}\,f_{\beta,N} \to \mathsf{E}\,f_\beta$ *and* $\mathsf{E}\,\widetilde{f}_{\beta,N} \to \mathsf{E}\,f_\beta$ *where*

$$\mathsf{E}\,f_\beta(q) = \min\left\{\beta^{-1}\sqrt{\frac{2\ln 2}{(\widehat{A})'(q)}}, 1\right\} = \begin{cases} \beta^{-1}\sqrt{\frac{2\ln\bar\alpha_j}{\bar a_j}} & \text{if } q \in [q_{j-1},q_j), j \leq l(\beta) \\ 1 & \text{if } q \geq q_{l(\beta)}. \end{cases} \tag{3.14}$$

The function f_β is a step function that jumps at points $\{0, q_1, \ldots, q_{l(\beta)}\}$. For any $q \in [q_{i-1}, q_i)$

$$f_\beta(q) = \int \mathcal{W}^{(l(\beta))}_{\beta\overrightarrow{\gamma}}(dw_1, \ldots, dw_{l(\beta)})w_{l(\beta)}(1 - w_i), \quad i = 1, \ldots, l(\beta); \quad (3.15)$$

$f_\beta(q) = 1$ for $q \geq q_{l(\beta)}$.

Rather then just considering the distribution of the total overlap, we can give a more precise description of the Gibbs measure by considering the vector of overlaps within each hierarchy. Let

$$\Delta_l = [-\ln \bar{\alpha}_l / \ln 2, \ln \bar{\alpha}_l / \ln 2], \quad \text{for } l = 1, 2, \ldots, l(\beta) \quad (3.16)$$
$$\Delta_{l(\beta)+1} = [-\ln(\alpha_{J_{l(\beta)}+1} \cdots \alpha_n)/\ln 2, \ln(\alpha_{J_{l(\beta)}+1} \cdots \alpha_n)/\ln 2].$$

It is clear that $(\bar{\sigma}_l \cdot \bar{\sigma}'_l)/N \in \Delta_l$, $l = 1, \ldots, l(\beta) + 1$. We introduce the measure $f^{\otimes l(\beta)+1}_{\beta,N}$ on $\Delta_1 \times \Delta_2 \times \cdots \times \Delta_{l(\beta)+1}$ induced by $\bar{\sigma}_l \cdot \bar{\sigma}'_l$ on all levels of limiting probability cascades: for any $I_l \in \Delta_l$, $l = 1, \ldots, l(\beta) + 1$, we put

$$f^{\otimes l(\beta)+1}_{\beta,N}(I_1 \times \cdots \times I_{l(\beta)+1}) \equiv \frac{\mathsf{E}_{\sigma,\sigma'} \prod_{l=1}^{l(\beta)+1} \mathbb{1}_{(\bar{\sigma}_l \cdot \bar{\sigma}'_l)/N \in I_l} e^{\beta\sqrt{N}(X_\sigma + X_{\sigma'})}}{Z^2_{\beta,N}}$$

$$= \mu^{\otimes 2}_{\beta,N} \left(\prod_{l=1}^{l(\beta)+1} \mathbb{1}_{(\bar{\sigma}_l \cdot \bar{\sigma}'_l)/N \in I_l} \right). \quad (3.17)$$

Theorem 3.4. [10] *The measure $f^{\otimes l(\beta)+1}_{\beta,N}$ converges in law to the following measure on $\Delta_1 \times \Delta_2 \times \cdots \times \Delta_{l(\beta)+1}$:*

$$f^{\otimes l(\beta)+1}_{\beta,N} \to Q_0 \delta_{(0,0,\ldots,0)} + \sum_{j=1}^{l(\beta)} Q_j \delta_{(\ln \bar{\alpha}_1 / \ln 2, \ldots, \ln \bar{\alpha}_j / \ln 2, 0, \ldots, 0)} \quad N \to \infty.$$

The random variables $Q_1, \ldots, Q_{l(\beta)}$ are defined as

$$Q_j(\beta) \equiv \int \mathcal{W}^{(l(\beta))}_{\beta\overrightarrow{\gamma}}(dw_1, \ldots, dw_{l(\beta)})w_{l(\beta)}(w_j - w_{j+1}\mathbb{1}_{\{j \leq l(\beta)-1\}}), \quad j = 1, \ldots, l(\beta).$$

3.4 Ghirlanda–Guerra Identities

The process $\mathcal{W}^{(l(\beta))}_{\beta\overrightarrow{\gamma}}$ has been constructed explicitly in Theorem 3.2 in terms of Ruelle's probability cascades. This allows to compute all its characteristics. Now we present a different approach that determines $\mathcal{W}^{(l(\beta))}_{\beta\overrightarrow{\gamma}}$ completely, without the use of Ruelle's probability cascades. This amounts to the computation of all moments of $\mathcal{W}^{(l(\beta))}_{\beta\overrightarrow{\gamma}}$ by recursion, starting from the second one. This approach will bear its full fruits in the analysis of the CREM.

Lemma 3.1. **[10]** *Then for any bounded function $h : \Sigma_N^n \to \mathbb{R}$ and for any $i = 1, \ldots, n$*

$$\left| \mathsf{E}\, \mu_{\beta,N}^{\otimes n+1}\left(h(\sigma^1, \ldots, \sigma^n)\mathbb{1}_{\sigma_1^k \ldots \sigma_i^k = \sigma_1^{n+1} \ldots \sigma_i^{n+1}}\right) \right.$$

$$\left. - \frac{1}{n}\, \mathsf{E}\, \mu_{\beta,N}^{\otimes n}\left(h(\sigma^1, \ldots, \sigma^n)\left(\sum_{l \neq k}^{n} \mathbb{1}_{\sigma_1^l \ldots \sigma_i^l = \sigma_1^k \ldots \sigma_i^k} + \mathsf{E}\, \mu_{\beta,N}^{\otimes 2}(\mathbb{1}_{\sigma_1^1 \ldots \sigma_i^1 = \sigma_1^2 \ldots \sigma_i^2})\right)\right) \right|$$

$$= \delta_N(a_i), \tag{3.18}$$

where $\delta_N(x) \geq 0$ here and in the sequel denotes a parameter such that, for any arbitrary small interval I,

$$\lim_{N \uparrow \infty} \int_I \delta_N(x)dx = 0. \tag{3.19}$$

The proof of this lemma is based on the integration by parts of Gaussian random variables coupled with a concentration of measure argument. The form of the error term asserts that, as N increases, the left-hand side vanishes eventually for almost all values of the parameters a_i, but it does not imply convergence for any fixed values. In the GREM, we could use the explicit convergence results for the partition function to deduce that we get actually convergence to zero for all but exceptional values of these parameters, but this is somewhat contrary to the spirit to this approach and will not work for the CREM or for SK models.

This lemma determines the so-called *Ghirlanda–Guerra identities* for the GREM: it allows to compute the expected distance distribution function between between n replicas under the Gibbs measure by the recurrence procedure (3.18) for $n = 3, 4, \ldots$ subsequently starting from $n = 2$. To see this, it suffices to put the function h equal to the indicator function of distances between $n + 1$ replicas and to note that by (3.18) the term with $n + 1$ replicas is completely determined by the terms with n replicas and by the one with two replicas

$$\mathsf{E}\, \mu_{\beta,N}^{\otimes 2}(\mathbb{1}_{\sigma_1^1 \ldots \sigma_i^1 = \sigma_1^2 \ldots \sigma_i^2}) = 1 - \mathsf{E}\, f_{\beta,N}\left(\sum_{j=1}^{i} \ln \alpha_j / \ln 2\right),$$

that has been already computed in (3.14). In fact, let $\underline{J} \equiv (\underline{J}_0, \ldots, \underline{J}_N)$ be a set of subsets of $1, \ldots, n + 1$ that determines the distances between $n + 1$ replicas: each element $\underline{J}_r = (J_{r,1}, \ldots, J_{r,j_r})$ is a collection of subsets of $1, \ldots, n + 1$ that reassembles the numbers of configurations for which the first r coordinated of the spin variables are equal. Then, for any $J_{r,i}$, there exists $J_{r-1,k}$, such that $J_{r,i} \subset J_{r-1,k}$. Assume that $J_{r,i}$ is the set of numbers $\{j_1^{r,i}, \ldots, j_{|J_{r,i}|}^{r,i}\}$. We can then define the function:

$$\mathcal{A}_{\underline{J}} \equiv \prod_{r=1}^{N} \prod_{i=1}^{j_r} \mathbb{1}_{\{\sigma_1^{j_1^{r,i}} \ldots \sigma_r^{j_1^{r,i}} = \cdots = \sigma_1^{j_{|J_{r,i}|}^{r,i}} \ldots \sigma_r^{j_{|J_{r,i}|}^{r,i}}\}}. \tag{3.20}$$

The length of $\underline{J}$ is $\|\underline{J}\| = n + 1$. Let us construct a set $\underline{J}'$ of length n by erasing everywhere in $\underline{J}$ the integer $n+1$. Indeed, there exists $r \in \{1, \ldots, N\}$, such that there exists $l \in \{1, \ldots, n\}$, such that $n + 1$ and l belong to the same subset, $J_{r,i}$, of $\underline{J}$, i.e. their first r coordinates coincide. If we choose the maximal r with this property, this determines uniquely the participation of $n+1$ everywhere in $\underline{J}$: for any $p = 1, 2, \ldots, r - 1$ it belongs to the same subset $J_{p,i}$ as l. In other words, once the ultrametric distances between n replicas are fixed, it suffices to specify the distance of the $(n + 1)$th replica to the closest to it, in order to determine completely its distance to all other replicas. This implies

$$A_{\underline{J}} = A_{\underline{J}'} \mathbb{1}_{\sigma_1^l \ldots \sigma_r^l = \sigma_1^{n+1} \ldots \sigma_r^{n+1}}. \tag{3.21}$$

Hence, substituting $h = A_{\underline{J}'}$ in Lemma 3.1, we can compute $\lim_{N \to \infty} \mathbb{E} \, \mu_{\beta,N}^{\otimes n}$ $(A_{\underline{J}})$ subsequently for $n = 3, 4, \ldots$, starting from $n = 2$, given by (3.14).

From the other hand, in [10], we expressed all moments of $\mathcal{W}_\beta^{(m)}$ in terms of $A_{\underline{J}}$:

$$\int \mathcal{W}_{\beta,N}^{(m)}(dw) w_1^{i_1} \ldots w_l^{i_l} \ldots w_m^{i_m}$$

$$= \mu_{\beta,N}^{\otimes(i_1 + \cdots + i_m)} \Big(\mathbb{1}_{\{\sigma_1^1 = \cdots = \sigma_1^{i_1 + \cdots + i_m}\}} \cdots \mathbb{1}_{\{\sigma_1^{i_1 + \cdots + i_{l-1}+1} \ldots \sigma_l^{i_1 + \cdots + i_{l-1}+1}}$$

$$= \cdots = \sigma_1^{i_1 + \cdots + i_m} \ldots \sigma_l^{i_1 + \cdots + i_m}\} \cdots \mathbb{1}_{\{\sigma_1^{i_1 + \cdots + i_{m-1}+1} \ldots \sigma_m^{i_1 + \cdots + i_{m-1}+1}}$$

$$= \cdots = \sigma_1^{i_1 + \cdots + i_m} \ldots \sigma_m^{i_1 + \cdots + i_m}\} \Big) \tag{3.22}$$

where $i_m \geq 1$, otherwise this expression is infinite. This implies the following theorem.

Theorem 3.5. [10] *The process* $\mathcal{W}_{\beta\vec{\gamma}}^{(m)}$ *is completely determined by the relations (3.18) up to the mean value of the two-replica distance distribution function given by (3.14).*

Theorem 3.5 in the case of the REM has been first proven by Talagrand. Lemma 3.1 implies also the following result, that has been remarked by Ruelle in [20].

Corollary 3.1. *The lth marginal of Ruelle's process of probability cascades* $\mathcal{W}_{\beta\vec{\gamma}}^{(m)}$ *with m levels and parameters $\beta\gamma_1 > \cdots > \beta\gamma_m > 1$ has the same distribution as Ruelle's process of one level with parameter $\beta\gamma_l$, $l = 1, \ldots, m$*

To see this, we need to control all moments of this marginal that can be expressed via the quantities $\mu_{\beta,N}^{\otimes r}\big(\mathbb{1}_{\{\sigma_1^1 \ldots \sigma_l^1 = \cdots = \sigma_1^r \ldots \sigma_l^r\}}\big)$, which in turn satisfy the identities (3.18) for $r = 3, \ldots$, while for $r = 2$ they are defined by $f_\beta(q_l) = (\beta\gamma_l)^{-1}$. But these identities are the same for the GREM with one hierarchy (i.e. the REM), with the same two replica distance distribution. Consequently, the lth marginal of $\mathcal{W}_{\beta,N}^{(m)}$, in the limit $N \to \infty$ behaves as $\sum_\sigma \delta_{\mu_{\beta,N}(\sigma)}$ of the REM at temperature $\tilde{\beta} = \beta\gamma_l \sqrt{2 \ln 2}$.

3.5 Empirical Distance Distribution Function $\mathcal{K}_\beta$

The process $\mathcal{W}_{\beta,N}^{(l(\beta))}$ is a point process on $[0,1]^m$. Its points $(\mu_{\beta,N}(B_1(\sigma)), \dots,$
$\mu_{\beta,N}(B_{l(\beta)}(\sigma)))$ can be considered as values of the ultrametric distance distribution function around σ

$$m_\sigma(x) = \mu_{\beta,N}(d_N(\sigma,\sigma') \le 1 - x) \tag{3.23}$$

at points $x = q_0, \dots, q_{l(\beta)}$. The limit of this distribution function is a step-function that jumps precisely at these points. We could then consider $\mathcal{W}_{\beta,N}^{(l(\beta))}$ as a point process of these distribution functions: $\mathcal{W}_{\beta,N}^{(l(\beta))} = \sum_\sigma \delta_{m_\sigma(\cdot)}$.

This object is, however, not properly adapted to the CREM. In the analysis of the CREM it is essentially imperative to replace it by the probability measure on these distribution functions:

$$\mathcal{K}_{\beta,N} = \sum_\sigma \mu_{\beta,N}(\sigma)\delta_{m_\sigma(\cdot)}, \tag{3.24}$$

that we have introduced and discussed in the introduction. To conclude the analysis of the GREM, we give its asymptotic behaviour in the following theorem.

Theorem 3.6. [31] *The process $\mathcal{K}_{\beta,N}$ converges weakly to the point process $\mathcal{K}_\beta$*

$$\mathcal{K}_\beta = \int_{\mathbb{R}^{l(\beta)}} \mathcal{W}_{\beta\overrightarrow{\gamma}}^{(l(\beta))}(dw)w(l(\beta))\delta_{m(w)} \tag{3.25}$$

where the measures $m(w)$ are defined by the formulas :

$$m(w) = (1 - w(1))\delta_1$$
$$+ (w(1) - w(2))\delta_{1-\ln\alpha_1/\ln 2} + \cdots + w(l(\beta))\delta_{1-\ln(\alpha_1\cdots\alpha_{l(\beta)})/\ln 2}. \tag{3.26}$$

4 CREM: Implicit Approach

We start now the analysis of the CREM with covariances (1.2) where $A(x) :$
$[0,1] \to [0,1]$ is a right-continuous distribution function with the concave hull $\widehat{A}(x)$ whose right derivative we denote by $(\widehat{A})'(x)$; see Fig. 5a. We assume that A is non-critical in the sense that it is equal to its concave hull $\widehat{A}$ only on the set of extremal points of the convex hull.

4.1 Maximum of the Hamiltonian. Limit of the Free Energy

Theorem 4.1. [11] *Let $\{X_\sigma\}$ be a family of 2^N standard Gaussian random variables with covariances (1.2). Then*

$$\lim_{N\to\infty} \mathsf{E}\max_\sigma \frac{X_\sigma}{\sqrt{N}} = \sqrt{2\ln 2}\int_0^1 \sqrt{(\widehat{A})'(x)}dx. \tag{4.1}$$

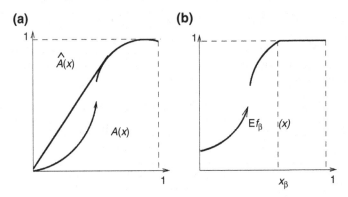

(a)   **(b)**

Fig. 5. (a) Concave hull of $A(x)$, (b) The function (4.5).

Theorem 4.2. [11] *Let $\{X_\sigma\}$ be a family of 2^N Gaussian random variables with covariances (1.2). Let*

$$x_\beta = \sup\left\{x \,\middle|\, (\widehat{A})'(x) > \frac{2\ln 2}{\beta^2}\right\}. \tag{4.2}$$

Then

$$\lim_{N\to\infty} N^{-1}\mathsf{E}\ln Z_{\beta,N} = \sqrt{2\ln 2}\beta \int_0^{x_\beta} \sqrt{(\widehat{A})'(x)}\,dx + \frac{\beta^2}{2}(1 - \widehat{A}(x_\beta)) + \ln 2(1 - x_\beta). \tag{4.3}$$

Consequently the critical temperature of the CREM defined as

$$\beta_0 = \sup\{\beta : \lim_{N\to\infty} N^{-1}\mathsf{E}\ln Z_{\beta,N} = \lim_{N\to\infty} N^{-1}\ln \mathsf{E}\,Z_{\beta,N}\}$$

equals:

$$\beta_0 = \sqrt{\frac{2\ln 2}{\lim_{x\downarrow 0}(\widehat{A})'(x)}}. \tag{4.4}$$

The proofs of these theorems rely heavily on results already obtained for the GREM. Namely, we approximate $A(x)$ from above and below by step-functions for which corresponding results have been already established in Theorem 2.3 (ii) and (3.3) in the study of the GREM. Then the results announced in Theorems 4.1 and 4.2 follow from theorems about the comparison of the mean values of convex or concave functions of Gaussian processes implied by the comparison of the covariances of these processes (see Theorem 3.1 in [32](Kahane's Theorem)).

We are not able to evaluate the fluctuations of the partition function of the CREM. We anticipate that they depend not only on $\widehat{A}(x)$ in view of the analysis of the maximum of branching Brownian motion by Bramson [33]. But (4.3) suffices to deduce the following very important result which is in the basis of the description of the CREM's Gibbs measure.

4.2 Two-Replicas Ultrametric Distance Distribution Function

Theorem 4.3. [11] *Let* $\{X_\sigma\}$ *be a family of* 2^N *Gaussian random variables with covariances (1.2). Let* x_β *be defined by (4.2). Then*

$$\lim_{N\to\infty} \mathsf{E}\,\mu_{\beta,N}^{\otimes 2}(d_N(\sigma,\sigma') \geq 1 - x) = \mathsf{E}\,f_\beta(x) = \begin{cases} \beta^{-1}\sqrt{\dfrac{2\ln 2}{(\widehat{A})'(x)}} & \text{if } x < x_\beta \\ 1 & \text{if } x \geq x_\beta \end{cases}$$

(4.5)

The function $\mathsf{E}\,f_\beta(x)$ is illustrated on Fig. 5b. Let us sketch the main points of the proof. The result of Theorem 4.2 allows to compute the limit of the free energy

$$\lim_{N\to\infty} N^{-1}\,\mathsf{E}\ln Z_{\beta,N}^u = F_\beta^u$$

for the CREM where the function $A(x)$ is slightly perturbed by a small parameter $u > 0$ in a neighbourhood of the point x. Next, using the integration by parts of Gaussian random variables, we show that the desired quantity $\lim_{N\to\infty} \mathsf{E}\,\mu_{\beta,N}^{\otimes 2}(d_N(\sigma,\sigma') \geq 1 - x)$ is equal to $\lim_{N\to\infty} \frac{d}{du} N^{-1} \ln Z_{\beta,N}^u \big|_{u=0}$ that can be computed as $\frac{d}{du} F_\beta^u \big|_{u=0}$ by convexity and leads to (4.5).

4.3 Ghirlanda–Guerra Identities

Lemma 4.1 is a generalisation of Lemma 3.1: it proves Ghirlanda–Guerra identities in the case of the CREM.

Lemma 4.1. [11] *For any* $n \in \mathbb{N}$, *any bounded function* $h(x)$ *and* $x \in [0,1] \setminus x_\beta$

$$\left| \mathsf{E}\,\mu_{\beta,N}^{\otimes n+1}\left(h(\sigma^1,\ldots,\sigma^n)\,\mathbb{I}_{d_N(\sigma^k,\sigma^{n+1})>x}\right) \right.$$

$$\left. -\frac{1}{n}\,\mathsf{E}\,\mu_{\beta,N}^{\otimes n+1}\left(h(\sigma^1,\ldots,\sigma^n)\left(\sum_{l\neq k}^{n} \mathbb{I}_{d_N(\sigma^k,\sigma^l)>x} + \mathsf{E}\,\mu_{\beta,N}^{\otimes 2}(\mathbb{I}_{d_N(\sigma^1,\sigma^2)>x})\right)\right) \right|$$

$$= \delta_N(u).$$

(4.6)

where this time the error term vanishes when we integrate over deformations of the function A *at the point* x.

One of the pillars of the proof of this lemma is the representation $X_\sigma = X_\sigma(1)$ where $X_\sigma(t)$ is the family of standard Gaussian processes on $[0,1]$ with covariances: $\mathrm{cov}\,(X_\sigma(t), X_{\sigma'}(s)) = A(t \wedge s \wedge d_N(\sigma,\sigma'))$. Two other pillars are the same as in the case of the GREM: the integration by parts of Gaussians and a concentration of measure argument.

This lemma implies the following important Theorem 4.4, that determines implicitly the empirical distance distribution function $\mathcal{K}_{\beta,N}$. Let us define a family of measures $\mathcal{Q}_N^n$ on $[0,1]^{n(n-1)/2}$

$$\mathcal{Q}_{\beta,N}^{(n)}(\bar{d}_N \in \mathcal{C}) \equiv \mathsf{E}\,\mu_{N,\beta}^{\otimes n}(\bar{d}_N \in \mathcal{C}) \tag{4.7}$$

where $\bar{d}_N$ is the vector of distances between n replicas with components $d_N^{k,l} = d_N(\sigma^l, \sigma^k)$, $1 \le l < k \le n$, and $\mathcal{C}$ is a Borel subset of $[0,1]^{n(n-1)/2}$. We denote by $\mathcal{B}_k$ the sigma-field generated by the first $k(k-1)/2$ coordinates.

Theorem 4.4. [11] *For any $n \in \mathbb{N}$, the family of measures $\mathcal{Q}_{\beta,N}^{(n)}$ converges,[4] as $N \uparrow \infty$, to the limiting measure $\mathcal{Q}_\beta^{(n)}$. All these measures are uniquely determined by (4.5). They satisfy the identities:*

$$\mathcal{Q}_\beta^{(n+1)}\big(d^{k,n+1} \in C|\mathcal{B}_n\big) = \frac{1}{n}\mathcal{Q}_\beta^{(2)}(C) + \frac{1}{n}\sum_{l=1,l\neq k}^{n} \mathcal{Q}_\beta^{(n)}\big(d^{k,l} \in C|\mathcal{B}_n\big) \tag{4.8}$$

for any Borel subset $C \subset [0,1]$. Consequently $\mathcal{K}_{\beta,N}$ defined by (3.24) and (3.23) converges in law to the limit $\mathcal{K}_\beta$ with generalized moments determined by $\mathcal{Q}_\beta^{(n)}$.

The recurrent formulas (4.8) come from (4.6) if we put h equal to the indicator function of any desired event of $\mathcal{B}_n$. Let us remark also that, due to the ultrametric structure, once the distances between n replicas are prescribed, it suffices to fix the distance from the $(n+1)$-th replica up to the closest to it among the n replicas $\{1, 2, \ldots, n\}$, in order to determine its distance up to all other $n-1$ replicas. This fact is already formally explained in (3.20). Then the formulas (4.8) determine completely the measures $\lim_{N\to\infty}\mathcal{Q}_{\beta,N}^{(n)} = \mathcal{Q}_\beta^{(n)}$ up to the measure $\mathcal{Q}_\beta^{(2)}$ already computed in (4.5). The moments $\mathcal{K}_{\beta,N}$ can be expressed in terms of the measures $\mathcal{Q}_{\beta,N}^{(n)}$. This implies the convergence of $\mathcal{K}_{\beta,N}$ to a limiting object $\mathcal{K}_\beta$ with moments expressed in terms of $\mathcal{Q}_\beta^{(n)}$.

4.4 Marginals of $\mathcal{K}_\beta$ in Terms of Ruelle's Probability Cascades

In this subsection we give an explicit form of all marginals of $\mathcal{K}_{\beta,N}$. Let $0 < t_1 < t_2 < \cdots < t_m < 1$ be points of increase of the function (4.5), $t_0 = 0$. We can define then the marginal process:

$$\mathcal{K}_{\beta,N}(t_0, t_1, \ldots, t_m) = \sum_\sigma \mu_{\beta,N}(\sigma)\delta_{m_\sigma(t_0), m_\sigma(t_1), m_\sigma(t_2), \ldots, m_\sigma(t_m)}. \tag{4.9}$$

Theorem 4.5. [11] *Let $t_0 = 0 < t_1 < t_2 < \ldots < t_m \le 1 = t_{m+1}$ be points of increase of the function (4.5). Consider the GREM of $m+1$ hierarchies, with parameters α_i such that $\ln\alpha_i/\ln 2 = t_i - t_{i-1}$, $i = 1, \ldots, m+1$, a_i*

---

[4] Here convergence is understood that we first average the N-dependent quantities with respect to deformations of size $\eta > 0$ of the functions A, then take the limit and finally take the limit $\eta \downarrow 0$.

with $\sum_{i=1}^{m+1} a_i = 1$ at temperature $\widetilde{\beta}$ such that $\widetilde{\beta}^{-1}\sqrt{2\ln\alpha_i/a_i} = \beta^{-1}\sqrt{2\ln 2/(\widehat{A})'(t_{i-1})}$, $i = 1, \ldots, m+1$. Then

$$\lim_{N\to\infty} \mathcal{K}_{\beta,N}(t_0, t_1, \ldots, t_m) = \mathcal{K}_{\widetilde{\beta}}^{(m+1)}, \tag{4.10}$$

where $\mathcal{K}_{\widetilde{\beta}}^{(m+1)}$ is the empirical distance distribution function of the GREM computed in Theorem 3.6 in terms of Ruelle's probability cascades.

The second moments of $\lim_{N\to\infty} \mathcal{K}_{\beta,N}(t_0, t_1, \ldots, t_m)$ and of $\mathcal{K}_{\widetilde{\beta}}^{(m)}$ for the GREM in question are the same due to the choice of the parameters of the GREM. Then all their moments coincide by the Ghirlanda–Guerra identities. The parameters a_i and $\widetilde{\beta}$ explicitly are equal to

$$a_i = \kappa \frac{\widehat{A}'(t_{i-1})\ln\alpha_i}{\ln 2}, \quad \widetilde{\beta} = \kappa^{-1/2}\beta, \quad \kappa = \left(\sum_{i=1}^{m+1} \frac{\widehat{A}'(t_{i-1})\ln\alpha_i}{\ln 2}\right)^{-1}, \quad i = 1, \ldots, m+1.$$

5 Genealogies and Neveu's Branching Process

5.1 Problems with the Explicit Description of Limiting Gibbs Measures

We obtained an implicit description of the limiting Gibbs measure of the CREM via recursive computation of all moments of $\mathcal{K}_\beta$. Nevertheless, we would like to identify explicitly a limiting measure to which our Gibbs measures converge and that encodes the full geometric information contained in $\mathcal{K}_\beta$. This is not immediately possible for the following reason. In [25] one of us proposed to describe the infinite volume limit of the Gibbs measure for the REM by considering the image of the hypercube Σ_N on $[0, 1]$ through the map $r_N : \Sigma_N \to (0, 1]$ (1.7). However, the definition of $\mathcal{K}_{\beta,N}$ involves masses of sets $\{\sigma' : d_N(\sigma, \sigma') < 1-t\}$. If we map such sets on the unit interval via r_N, we obtain intervals $(r_{[Nt]} - 2^{-[tN]}, r_{[Nt]}]$ of length $2^{-[tN]}$. So, when $N = \infty$, these sets map to intervals of length $2^{-t\infty}$. We can not analyse the structure of the measure by looking at intervals of the size $2^{-t\infty}$.

What will however be possible, is the following. We will introduce the notion of a flow of compatible probability measures on $[0, 1]$ indexed by pairs of parameters $s \leq t \in I$ and with distribution functions satisfying the compatibility assumption (5.1). Next, we will associate to each of such flows a certain genealogical structure on $[0, 1]$ described by a genealogical map, $K_T \in M_1(M_1([0, 1]))$, which is an empirical distribution of family sizes of all individuals as functions of degree of relatedness. Then we will provide a flow of compatible probability measures for each finite N with the genealogy describing efficiently the geometry of the Gibbs measure of the CREM: its genealogical map, $K_T^{\beta,N}$, will equal the empirical distance distribution function $\mathcal{K}_{\beta,N}$. Finally, we will show that this flow of probability measures converges as $N \to \infty$ to the flow of compatible random probability measures with distribution functions that are normalised stable subordinators associated to Neveu's continuous state branching process via an appropriate deterministic time change. This convergence of flows is understood in the sense that their genealogical maps, $\mathcal{K}_{\beta,N} = K_T^{\beta,N}$, converge. Thus, the limiting geometry of the Gibbs measure of the CREM will be expressed in terms of the genealogy of Neveu's continuous state branching process modulo a time change determined only by $\mathsf{E}\, f_\beta(x)$ of (4.5).

5.2 Genealogical Map of a Flow of Probability Measures

Definition 5.1. *A two-parameter family of measures with probability distribution functions $S^{(s,t)}$ on $[0,1]$, $s \leq t$, $s,t \in I \subset \mathbb{R}$, is called a flow of compatible probability measures on I, if and only if for any collection $t_1 \leq t_2 \leq \cdots \leq t_n \subset I$*

$$S^{(t_1,t_n)} = S^{(t_{n-1},t_n)} \circ S^{(t_{n-2},t_{n-1})} \circ \ldots S^{(t_2,t_3)} \circ S^{(t_1,t_2)} \tag{5.1}$$

holds.

Let us admit the following terminology. We say that each point $a \in [0,1]$ is an individual in generation s and its image $S^{(s,t)}(a) \in [0,1]$ is its offspring in generation t. Let us define for any distribution function $\Theta(x)$ its inverse function

$$\Theta^{-1}(x) = \inf\{a \mid \Theta(a) \geq x\}. \tag{5.2}$$

Then each individual $x \in [0,1]$ in generation t has an ancestor a in generation s which is $a = (S^{(s,t)})^{-1}(x)$. Given an individual $x \in [0,1]$ in generation t, let us look for individuals x' having the same ancestor as x in generation s

$$m_x(s,t) \equiv \{x' : (S^{(s,t)})^{-1}(x') = (S^{(s,t)})^{-1}(x)\}. \tag{5.3}$$

If $S^{(s,t)}$ is continuous at $a = (S^{(s,t)})^{-1}(x)$, then any individual $x' \neq x$ has a different ancestor from the one of x. If $S^{(s,t)}$ makes a jump at $a = (S^{(s,t)})^{-1}(x)$, then the family (5.3) of the individual x having the same ancestor as x in generation s is the following interval:

$$m_x(s,t) = \lim_{\eta \downarrow 0} \left(S^{(s,t)}\big((S^{(s,t)})^{-1}(x) - \epsilon\big),\ S^{(s,t)} \circ (S^{(s,t)})^{-1}(x) \right].$$

In Fig. 6 the individual x in generation t has a family of "cousins" $m_x(s,t)$ having the same "grand-father" in generation s, while the individual y is the unique "grand-child" of his ancestor in generation s. We are mainly interested in a non-trivial case when functions $S^{(s,t)}$ make jumps. The next lemma

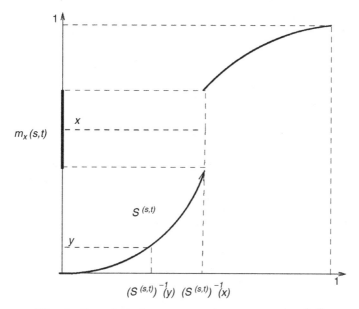

Fig. 6. Genealogical structure induced by a flow $S^{(s,t)}$

justifies this terminology. It says that any individual having an ancestor in common with x in generation s has necessarily an ancestor in common with x in any generation $s' < s$. In other words, if we partition the interval $[0,1]$ into families $m_x(s',t)$ having the same ancestor in generation s', then the partition into families $m_x(s,t)$ having the same ancestor in generation $s > s'$ is a refinement of the previous one.

Lemma 5.1. *Let $S^{(s,t)}$ be distribution functions of a flow of measures according to Definition 5.1. Then for all $x \in [0,1]$*

$$m_x(s,t) \subset m_x(s',t) \quad \forall s' < s \le t \in I. \tag{5.4}$$

Whenever $t = T$ is fixed, the function $|m_x(\cdot,T)|$ is the family size of the individual x in generation T as a function of the degree of relatedness. By Lemma 5.1, it is a decreasing function on I. Finally, we define the associated empirical distribution of the functions $|m_x(\cdot,T)|$

$$K_T = \int\limits_0^1 dx\, \delta_{|m_x(\cdot,T)|}. \tag{5.5}$$

This construction allows to associate to any flow of probability measures, in the sense of Definition 5.1, an empirical distribution K_T. If we assume, in addition, that $[0,T] \subset I$ and $|m_x(\cdot,T)|$ are right-continuous, then $1 - |m_x(\cdot,T)|$ are probability distribution functions. Then we will think of K_T as a map from flows of probability measures into $M_1(M_1([0,1]))$ which we call the genealogical map.

5.3 Coalescent Associated with a Flow of Probability Measures

Now, let us define the exact degree of relatedness between two individuals $x, y \in [0, 1]$ with respect to a flow of measures (5.1) as

$$\gamma_T(x, y) \equiv \sup \left(s \in I : y \in m_x(s, T) \right). \tag{5.6}$$

Lemma 5.2. $T - \gamma_T$ *defines an ultrametric distance on the unit interval.*

We will be interested in cases where the flow $S^{(s,t)}$ of Definition 5.1 is *random*. We will now define the coalescent process on integers that completely characterises a random genealogical map K_T in this case.

Having defined a distance $T - \gamma_T$ on $[0, 1]$, we can define in a very natural way the analogous distance on the integers. To do this, consider a family of i.i.d. random variables, $\{U_i\}_{i \in \mathbb{N}}$, distributed according to the uniform law on $[0, 1]$. Given such a family, we set

$$\rho_T(i, j) = \rho_T(U_i, U_j). \tag{5.7}$$

Due to the ultrametric property of the ρ_T and the independence of the U_i, for fixed T, the sets $B_i(s) \equiv \{j : \rho_T(i, j) \leq T - s\}$ form an exchangeable random partition of the integers. Moreover, the family of these partitions as a function of $T - s$ is a stochastic process on the space of integer partitions with the property that for any $s > s'$, the partition $B_i(s')$ is a coarsening of the partition $B_i(s)$. Such a process is called a *coalescent process*.

The key observation is the following lemma.

Lemma 5.3. *The genealogical map K_T of a flow $S^{(s,t)}$ is completely determined by its moments; they can be expressed through the probabilities*

$$\mathsf{P}(\rho_T(i, j) \leq T - t_{m(i,j)}, \;\; \forall i, j \in \{1, \ldots, l\}) \tag{5.8}$$

of the corresponding coalescent, where $m(i, j) \in \{1, \ldots, p\}$, $0 < t_1 < \cdots < t_p \leq T$, $l \geq 2$.

To illustrate this lemma, let us note that

$$\mathsf{E} \int m(t) K_T(dm) = \mathsf{E} \int_0^1 m_x(T, t) dx = \mathsf{P}(\rho_T(1, 2) \leq T - t). \tag{5.9}$$

5.4 Finite N Setting for the CREM

We will now show that for finite N we can use the general construction from Sects. 5.2 and 5.3 to relate the geometric description of the Gibbs measure on Σ_N to the genealogical description of a family of embedded measures on $[0, 1]$.

Recall that we have already introduced the image measure $\widetilde{\mu}_{\beta,N}$ (1.8) of the Gibbs measure on the unit interval via the map r_N (1.7). Let $\theta_{\beta,N}$ be the probability distribution function of $\widetilde{\mu}_{\beta,N}$:

$$\theta_{\beta,N}(x) = \widetilde{\mu}_{\beta,N}(\sigma : r_{[N]}(\sigma) \le x). \tag{5.10}$$

Let us take a parameter $s \in [0,1]$ and consider the map $r_{[sN]} : \Sigma_N \to [0,1]$. Clearly, its image consists of $2^{[sN]}$ points, and for any σ, σ' with $d_N(\sigma, \sigma') > s$ we have $r_{[sN]}(\sigma) = r_{[sN]}(\sigma')$. Now we define a family of compatible distribution functions in the sense of Definition 5.1

$$S_{\beta,N}^{(s,t)}(a) = \sum_{\sigma} \mu_{\beta,N}(\sigma) \, \mathbb{1}_{\{\theta(r_{[sN]}(\sigma)) \le a\}} \tag{5.11}$$

as states Lemma 5.4. To better understand the construction of (5.11), let us take configurations, $\sigma^1, \sigma^2, \ldots, \sigma^{2^{[sN]}}$, differing in the first $[sN]$ coordinates, i.e. with $d_N(\sigma^i, \sigma^j) \ge 1 - s$, and arrange them in order such that $0 < r_{[sN]}(\sigma^1) < r_{[sN]}(\sigma^2) < \cdots < r_{[sN]}(\sigma^{2^{[sN]}}) = 1$. Let

$$x_i^s = \mu_{\beta,N}(\sigma' : d_N(\sigma', \sigma^i) < 1 - s), \quad i = 1, \ldots, 2^{[sN]}, \quad x_0 = 0.$$

Define

$$y_i^s \equiv x_0^s + x_1^s + \cdots + x_i^s = \theta(r_{[sN]}(\sigma^i)), \quad i = 0, 1, \ldots, 2^{[sN]}.$$

Then we may write the representation

$$S_{\beta,N}^{(s,t)}(a) = \sum_{i=0}^{2^{[sN]}} y_i^s \, \mathbb{1}_{\{a \in [y_i^s, y_{i+1}^s)\}}. \tag{5.12}$$

Lemma 5.4. *The functions, $S_{\beta,N}^{(s,t)}$, defined in (5.11) satisfy the assumptions of Definition 5.1 with $I = [0,1]$.*

It follows from this observation that we are entitled to apply the construction of the previous section to $S_{\beta,N}^{(s,t)}$. Their genealogy is

$$m_x(s,t) = (y_{i-1}^s, y_i^s] \text{ with } |m_x(s,t)| = |x_i^s|, \text{ if } x \in (y_{i-1}^s, y_i^s], \; i = 1, \ldots, 2^{[sN]}.$$

We may associate with this genealogy the genealogical map, K_T, and the coalescent process on the integers. Lemma 5.5 expresses the geometry of the Gibbs measure of the CREM contained in the empirical distance distribution function $\mathcal{K}_{\beta,N}$, defined in (1.13), in terms of the genealogy induced by the functions defined in (5.11).

Lemma 5.5. *We have*

$$\mathcal{K}_{\beta,N} = K_1^{\beta,N},$$

where the empirical distance distribution function $\mathcal{K}_{\beta,N}$ is defined in (1.13) and $K_1^{\beta,N}$ is the genealogical map defined in (5.5), with $T = 1$, of the flow of probability distribution functions (5.11).

5.5 Genealogy of a Continuous State Branching Process

Another example of flows of probability measures satisfying Definition 5.1 arises in the context of continuous state branching process [23]. The basic object here is a continuous state branching process $X(t)$ on $\mathbb{R}^+$ characterised by its Laplace exponent $u_t(\lambda)$. The process started in $a \geq 0$ will be denoted by $X(\cdot, a)$. This can be extended to a genuine two parameter process $(X(t, a), t, a \geq 0)$ using the fundamental branching property that states that, if $X'(\cdot, b)$ and $X(\cdot, a)$ are independent copies, then $X(\cdot, a + b)$ has the same law as $X'(\cdot, b) + X(\cdot, a)$. The process $X(t, a)$ is characterised by the property that, for any $a, b \geq 0$, $X(\cdot, a + b) - X(\cdot, a)$ is independent of the processes $X(\cdot, c)$, for all $c \leq a$, and its law is the same as that of $X(\cdot, b)$. The right continuous version of $X(t, \cdot)$ is a subordinator. Bertoin and Le Gall [23] prove the following proposition, based on the Markov property of this process.

Proposition 5.1. *On some probability space there exists a process* $(\widetilde{S}^{(s,t)}(a)$, $0 \leq s \leq t, a \geq 0)$, *such that*
(i) For any $0 \leq s \leq t$, $\widetilde{S}^{(s,t)}$ *is a subordinator with Laplace exponent* $u_{t-s}(\lambda)$.
(ii) For any integer $p \geq 3$ *and* $0 \leq t_1 \leq t_2 \leq \cdots \leq t_p$, *the subordinators* $\widetilde{S}^{(t_1,t_2)}, \widetilde{S}^{(t_2,t_3)}, \ldots, \widetilde{S}^{(t_{p-1},t_p)}$ *are independent, and*

$$\widetilde{S}^{(t_1,t_p)}(a) = \widetilde{S}^{(t_{p-1},t_p)} \circ \widetilde{S}^{(t_{p-1},t_p)} \circ \ldots \circ \widetilde{S}^{(t_2,t_3)} \circ \widetilde{S}^{(t_1,t_2)}(a), \quad \forall a \geq 0, \text{ a.s.}$$
(5.13)

(iii) The processes $\widetilde{S}^{(0,t)}(a)$ *and* $X(t, a)$ *have the same finite dimensional marginals.*

The process $\widetilde{S}^{(s,t)}$ allows to construct a flow of probability distribution functions by setting

$$S^{(s,t)}(x) \equiv \frac{1}{X(t,1)} \widetilde{S}^{(s,t)}(X(s,1)x), \quad 0 \leq s \leq t \leq 1.$$
(5.14)

For I taken as any countable subset of $\mathbb{R}^+$, they satisfy the assumptions of Definition 5.1 a.s.

We are interested in a particular case of Neveu's continuous state branching process X_t with

$$E(e^{-\lambda X_t} \mid X_0 = a) = e^{-u_t(\lambda)a}, \quad u_t(\lambda) = \lambda^{e^{-t}}.$$
(5.15)

In this case $\widetilde{S}^{(s,t)}$ are *stable subordinators* with index e^{s-t}. Then the normalised stable subordinators $S^{(s,t)}$ of (5.14) is a family of random probability distribution functions satisfying Definition 5.1. Thus, the genealogical construction of Sects. 5.2 and 5.3 applies to them.

Finally, note that if we take an increasing function $t(y) \geq 0$ for $y \in [0, 1]$, then we may consider the time-changed flow $\bar{S}^{(y,z)} = S^{(t(y),t(z))}$, $0 \leq y \leq z$, satisfying again Definition 5.1 and therefore allowing the genealogical construction of Sects. 4.2 and 4.3

Bertoin and Le Gall [23] showed that the coalescent process on the integers induced by $S^{(s,t)}$ of (5.14) associated to Neveu's process (5.15) coincides with the coalescent process constructed by Bolthausen and Sznitman [22]. They also proved the following remarkable result connecting the collection of subordinators to Ruelle's Generalised Random Energy Model: Take the parameters $0 < x_1 < \cdots < x_p < 1$ and $0 < t_1 < \cdots < t_p$ linked by the identities

$$t_k = \ln x_{k+1} - \ln x_1 \tag{5.16}$$

for $k = 0, \ldots, p-1$, and $t_p = -\ln x_1$. Then the law of the family of jumps of the normalised subordinators $S^{(t_k, t_p)}$, for $k = 0, \ldots, p-1$, is the same as the law of Ruelle's probability cascades $\mathcal{W}^{(p)}$ with parameters x_i, $i = 1, \ldots, p$, see Definition 3.1..

Now consider a GREM with finitely many hierarchies and parameters such that the points $y_0 = 0$ and $0 < y_1 < \ldots < y_p \leq 1$ are the extremal points of the concave hull of A. Recall that $\lim_{N \to \infty} \mathsf{E} f_{\beta,N}(y) = \mathsf{E} f_\beta(y)$ can be computed by (4.5) for any $y \in [0, 1]$. Now set

$$\mathsf{E} f_\beta(y_{i-1}) = x_i, \quad i = 1, \ldots, p, \tag{5.17}$$

where all of the $x_i < 1$. In Theorem 3.2 we proved that the point process $\mathcal{W}_{N,\beta}^{(p)}$ in $[0,1]^p$ converge to Ruelle's probability cascades with parameters x_i, $i = 1, \ldots, p$. (The convergence of the marginals of the process $\mathcal{W}_{N,\beta}^{(p)}$ for the GREM under the assumption that for any given hierarchy $i = 1, \ldots, p$ and $N > 0$ the number of configurations $\{\sigma' : d_N(\sigma, \sigma') < 1 - y_i\}$ is the same for all $\sigma \in \Sigma_N$, has been also established in Proposition 9.6 of [26].) Combining these two results yields.

Lemma 5.6. *Let $\mu_{\beta,N}$ be the Gibbs measure associated to a GREM with finitely many hierarchies satisfying (5.17) at the extremal points y_i, $i = 1, \ldots, p$ of the concave hull of the function A. Then the family of distribution functions $S_{\beta,N}^{(y_k, y_p)}$, $k = 1, 2, \ldots, p$ defined according to (5.11) converges in law, and the limit has the same distribution as the family of normalised stable subordinators (5.14) $S^{(t_k, t_p)}$, $k = 0, 1, \ldots, p-1$ in the sense that the joint distribution of their jumps has the same law, provided t_k are chosen according to (5.16), (5.17).*

5.6 Main Result

From the preceding proposition we expect that Neveu's process will provide the universal limit for all of our CREMs. The dependence on the particular model (i.e. the function A) and on the temperature must come from a rescaling of time. Set

$$x(y) \equiv \mathsf{E} f_\beta(y) = \begin{cases} \dfrac{\sqrt{2 \ln 2}}{\beta \sqrt{\widehat{A}'(y)}}, & \text{if } y < y_\beta \\ 1, & \text{if } y \geq y_\beta, \end{cases} \tag{5.18}$$

where $y_\beta = \sup(y : \frac{\sqrt{2\ln 2}}{\beta\sqrt{\widehat{A}'(y)}} < 1)$ (here $\mathsf{E}\, f_\beta(y)$ is defined by the function A through (4.5)). Set also

$$T = -\ln x(0), \quad t(y) = T + \ln x(y). \tag{5.19}$$

Define the flow of probability distribution functions

$$\bar{S}^{(y,z)}(x) \equiv S^{(t(y),t(z))}(x), \tag{5.20}$$

where $S^{(s,t)}$ is the flow of functions (5.14) associated to Neveu's process (5.15). Let $\bar{K}_T^{t(y)}$ be the genealogical map (5.5) associated to this flow.

Theorem 5.1. *Then*

$$\mathcal{K}_{\beta,N} = K_1^{\beta,N} \xrightarrow{\mathcal{D}} \bar{K}_1^{t(y)}. \tag{5.21}$$

Here $\mathcal{K}_{\beta,N}$ is the empirical distance distribution function (1.13), $K_1^{\beta,N}$ is the genealogical map (5.5) of the flow of probability distribution functions (5.11) and the equality $\mathcal{K}_{\beta,N} = K_1^{\beta,N}$ holds by Lemma 5.5. Theorem 5.1 is the main result of this paper. It expresses the geometry of the limiting Gibbs measure contained in $\mathcal{K}_{\beta,N}$ in terms of the genealogy of Neveu's branching process via the deterministic time change (5.19). We prove this theorem in Sect. 5.7.

5.7 Coalescence and Ghirlanda–Guerra identities

As it was remarked in Sect. 5.3, K_T associated with a flow of measures is completely determined by its moments, and these can be expressed via genealogical distance distributions of the corresponding coalescent (5.8). So, we will prove that the moments of $\mathcal{K}_{\beta,N}$, which are the n-replica distance distributions in our spin-glass model, converge to the genealogical distance distributions on the integers (5.8) constructed from the flow of compatible measures with distribution functions $\bar{S}^{(y,z)}$ (5.20). But the flow $\bar{S}^{(y,z)}$ is the time-changed flow (5.14) of Neveu's branching process (5.15) that by [23] corresponds to the coalescent of Bolthausen–Sznitman. Therefore, its genealogical distance distributions on the integers are those of Bolthausen–Sznitman coalescent under this time change (5.19). Then the proof of Theorem 5.1 is reduced to Theorem 5.2, which gives in addition the connection between the n-replica distance distribution function of the CREM with the genealogical distance distribution function of the Bolthausen–Sznitman coalescent.

Theorem 5.2. *Under the same assumptions as in Theorem 5.1, for any* $n \in \mathbb{N}$,

$$\lim_{N\uparrow\infty} \mathsf{E}\,\mu_{\beta,N}^{\otimes n}(d_N(\sigma^1,\sigma^2) \geq 1 - y_1,,\ldots,d_N(\sigma^{n-1},\sigma^n) \geq 1 - y_{n(n-1)/2}) \tag{5.22}$$

$$= \mathsf{P}\left(\rho_T(1,2) \geq T - t(y_1),\ldots,\rho_T(n-1,n) \geq T - t(y_{n(n-1)/2})\right)$$

where $t(y)$ is defined in (5.19) via (5.18).

The distance ρ_T is the distance on integers for the Bolthausen–Sznitman coalescent, induced through (5.7) by the genealogical distance γ_T of the flow of measures $S^{(s,t)}$ (5.14) of Neveu's branching process (5.15). The fact that in Bolthausen–Sznitman coalescent $\mathsf{P}(\rho_T(1,2) \geq T - t) = e^{t-T}$ and the convergence (4.5) imply the statement of the theorem for $n = 2$:

$$\mathsf{E}\,\mu_{\beta,N}^{\otimes 2}(d_N(\sigma,\sigma') \geq 1 - y) \to x(y) = e^{t(y)-T} = \mathsf{P}(\rho_T(1,2) \geq T - t(y)).$$

The proof of the theorem for $n > 2$, and in fact the entire identification of the limiting processes with objects constructed from Neveu's branching process, relies on the Ghirlanda–Guerra identities [28] that were derived in Theorem 4.4 for the left-hand side of (5.22). Thus we must show that the right-hand side of (5.22) satisfies the same identities, that is for $t < T$:

$$\mathsf{P}\left(\rho_T(1,n+1) \geq T - t \mid \mathcal{B}_n\right) = \frac{1}{n}e^{t-T} + \frac{1}{n}\sum_{k=2}^{n}\mathsf{P}\left(\rho_T(1,k) \geq T - t \mid \mathcal{B}_n\right)$$

$$(5.23)$$

that can be equivalently written as

$$\mathsf{P}\left(\rho_T(k,n+1) < T - t \mid \mathcal{B}_n\right) = \frac{|l \in \{1,\ldots,n\} : \rho_T(k,l) < T - t| - e^{t-T}}{n}$$

$$(5.24)$$

There are *two* ways to verify that (5.23) holds for the Bolthausen–Sznitman coalescent.

The first one is to observe that relation (5.23) involves only the marginals of the coalescent at a finite set of times. By Theorem 5 of Bertoin–Le Gall [23], these can be expressed in terms of Ruelle's probability cascades modulo the appropriate time change. Thus, by Theorem 3.2 these probabilities can be expressed as limits of a suitably constructed GREM (with finitely many hierarchies) for which the Ghirlanda–Guerra relations do hold by Lemma 3.1. Thus (5.23) is satisfied.

The second way is to verify directly that Ghirlanda–Guerra relations (5.24) hold for the Bolthausen–Sznitman coalescent.

This can be done by identifying its partitions with exchangeable random partitions called "Chinese restaurant process".

For that purpose, let us first give the following definition. Given the sequence of normalised jumps of the stable subordinator (Δ_i/T) with index x and given $U_1, U_2, \ldots$ independent uniform random variables on $[0, 1]$, the partition of positive integers Π distributed as a partition of blocks of indices of U_i belonging to the same intervals $\Delta_i/T \in [0, 1]$ is called $(x, 0)$-*partition*, see [34].

Let us introduce an operation of coagulation on partitions, see [35]: for a partition $\pi = (A_1, A_2, \ldots,)$ and $\Pi = (B_1, B_2, \ldots)$, the Π-coagulation of π consists of blocks of the form $\bigcup_{j \in B_i} A_j$.

By [22] *the Markov kernels* $(e^{-t}, 0)$-*coagulation,* $t \geq 0$, *on partitions of* $\mathcal{N}$ *form a semi-group. The Markov process*

$$\mathsf{P}^\pi(\Pi(t+) \in \cdot) = (e^{t-T}, 0) - coagulation\ of\ \pi \qquad (5.25)$$

is distributed as the Bolthausen–Sznitman coalescent. It starts from a partition of singletons at time T and finishes by a partition of one block $\mathcal{N}$ at time $-\infty$. (The semi-group property can be also seen from the fact that the limiting frequencies of $(e^{-t}, 0)$-partitions are distributed as normalised jumps of stable subordinators and from their matching condition (5.1).)

Next, consider exchangeable random partitions Π on $\mathbb{N}$, introduced by Pitman under the name of Chinese restaurant processes. For each parameter $0 < x < 1$ the partition called "Chinese restaurant process" can be constructed as follows. Let Π_n denote the restriction of Π to the first n positive integers. Then, conditionally given $\Pi_n = \{A_1, \ldots, A_k\}$ for any particular partition of $\{1, 2, \ldots, n\}$ into k subsets (tables) A_i of sizes n_i, $i = 1, \ldots, k$, the partition Π_{n+1} is an extension of Π_n such that the number $n + 1$ (new customer) is attached to the class (table) A_i with probability $(n_i - x)/n$, and forms a new class (sits at a new table) with probability kx/n. Let us denote by $p(n_1, \ldots, n_k)$ the probability of partitions Π with Π_n a particular partition of k classes of sizes $n_1, \ldots, n_k$, respectively. Then

$$p(n_1 + 1, n_2, \ldots, n_k) = \frac{n_1 - x}{n} p(n_1, \ldots, n_k). \qquad (5.26)$$

The crucial fact is that *the partition* Π *of the Chinese restaurant process with parameter* x *is a* $(x, 0)$-*partition.* This fact, noticed in [34], follows from the combination of the results of [35] and [36] : On the one hand, in [35] it is proven that the limiting relative frequencies, in order of appearance, P_i, in the Chinese restaurant process have the same distribution as the product $(1 - W_1)(1 - W_2) \cdots (1 - W_{i-1})W_i$, with W_i independent beta random variables with parameters $(1 - x, ix)$. On the other hand, in [36] the following was proven: let $\Delta_{(i)}/T$ denote the reordering to the intervals Δ_i/T in order of appearance of the U_i, i.e. define $\Delta_{(i)}$ such that $U_1 \in \Delta_{(1)}/T$, $U_{\min\{j : U_j \notin \Delta_{(1)}/T\}} \in \Delta_{(2)}/T$, etc., Then $|\Delta_{(i)}/T|$ has the same distribution as products, $(1 - W_1) \ldots (1 - W_{i-1})W_i$, where W_i are the independent beta random variables appearing above. Thus, the sequences $|\Delta_{(i)}/T|$ and P_i have the same distribution.

Therefore, by (5.25), the marginals of Bolthausen–Sznitman coalescent $\Pi(t)$ at times $0 = t_0 < t_1 < \cdots < t_{p-1} < t_p = T$ can be constructed as the following sequence of Chinese restaurant processes: let $x_i = e^{t_{i-1}-t_p}$, $0 < x_1 < x_2 < \cdots < x_p < 1$. Then $\Pi(t_{p-1}+)$ is distributed as a $(x_p, 0)$-partition, i.e. as the Chinese restaurant process with parameter x_p. Next, we define the partition $\Pi(t_{p-2}+)$ as the Chinese restaurant process on the classes of partition $\Pi(t_{p-1}+)$ with parameter $x_{p-1}/x_p = e^{t_{p-2}-t_{p-1}}$; this means that, given the classes $A_1^{p-1}, \ldots, A_k^{p-1}$ obtained from $A_1^p, \ldots, A_l^p$, where A_i^{p-1}

consists of l_i blocks of Π^p, $i = 1, \ldots, k$, $l_1 + \cdots + l_k = l$, the block A_{l+1}^p joins A_i^{p-1} with probability $(l_i^{p-1} - x_{p-1}/x_p)/l$ and forms a new class with probability $kx_{p-1}/(x_p l)$. One iterates this procedure with parameters $x_{p-2}/x_{p-1}, \ldots, x_1/x_2$ to construct the partitions $\Pi(t_{p-3}+), \ldots, \Pi(t_0+)$. By the semi-group property of $(e^{-t}, 0)$-coagulations, $\Pi(t_i+)$ is distributed as a Chinese restaurant process with parameter $x_{i+1} = e^{t_i - t_p}$ for all $i = 0, 1, \ldots, p - 1$, satisfying (5.26). Now (5.24) is immediate from the Chinese restaurant property (5.26).

Acknowledgement

A.B. is partially supported by the DFG in the Dutch–German bilateral research group "Mathematical models from physics and biology". We acknowledge support from the European Science Foundation though the programme RDSES.

References

1. F. Guerra, Broken replica symmetry bounds in the mean field spin-glass model, *Commun. Math. Phys.* **233**, 1–12 (2003)
2. M. Aizenman, R. Sims, and S.L. Starr. An extended variational principle for the SK spin-glass model. *Phys. Rev. B*, **6821**, 4403 (2003)
3. M. Talagrand, The generalized Parisi formula. *C. R. Math. Acad. Sci. Paris* **337**, 111–114 (2003)
4. M. Talagrand, The Parisi formula. to appear in *Ann. Math.* **163**, 221–263 (2006)
5. M. Talagrand, Self organization in the low-temperature region of a spin-glass model, *Rev. Math. Phys.* **15**, 1–78 (2003)
6. M. Talagrand, Spin-Glasses: A Challenge to Mathematicians, Ergebnisse der Mathematik und ihrer Grenzgebiete, vol. 46, Springer, Berlin Heidelberg New York (2003)
7. B. Derrida. A generalisation of the random energy model that includes correlations between the energies. *J. Phys. Lett.* **46**, 401–407 (1985)
8. B. Derrida and E. Gardner. Solution of the generalised random energy model. *J. Phys. C* **19**, 2253–2274 (1986)
9. B. Derrida and E. Gardner. Magnetic properties and function $q(x)$ of the generalised random energy model. *J. Phys. C* **19**, 5783–5798 (1986)
10. A. Bovier and I. Kurkova. Derrida's generalised random energy models 1: Poisson cascades and Gibbs measures. *Ann. de l'I.H.P.* **40**, 439–480 (2004)
11. A. Bovier and I. Kurkova. Derrida's generalised random energy models 2: Models with continuous hierarchies. *Ann. de l'I.H.P.* **40**, 481–495 (2004)
12. A. Bovier and I. Kurkova. Gibbs measures of Derrida's generalised random energy models and the genealogy of Neveu's continuous state branching process. Preprint, University Paris 6 (2003)
13. D. Sherrington and S. Kirkpatrick. Solvable model of a spin-glass. *Phys. Rev. Lett.* **35**, 1792–1796 (1972)

14. B. Derrida. Random energy model: Limit of a family of disordered models, *Phys. Rev. Letts.* **45**, 79–82 (1980)
15. B. Derrida. Random energy model: An exactly solvable model of disordered systems, *Phys. Rev. B* **24**, 2613–2626 (1981)
16. E. Gardner and B. Derrida. The probability distribution of the partition function of the random energy model. *J. Phys. A* **22**, 1975–1981 (1989)
17. D. Capocaccia, M. Cassandro, and P. Picco. On the existence of thermodynamics for the generalised random energy models. *J. Stat. Phys.* **46**, 493–505 (1987)
18. T.H. Eisele. On a third order phase transition, *Commun. Math. Phys.* **90**, 125–159 (1983)
19. A. Galvez, S. Martinez, and P. Picco. Fluctuations in Derrida's random energy and generalised random energy models. *J. Stat. Phys.* **54**, 515–529 (1989)
20. D. Ruelle. A mathematical reformulation of Derrida's REM and GREM. *Commun. Math. Phys.* **108**, 225–239 (1987)
21. J. Neveu. A continuous state branching process in relation with the GREM model of spin-glass theory. Rapport interne no. 267, Ecole Polytechnique
22. E. Bolthausen and A.S. Sznitman. On Ruelle's probability cascades and an abstract cavity method. *Commun. Math. Phys.* **107**, 247–276 (1998)
23. J. Bertoin and J.F. Le Gall. The Bolthausen–Sznitman coalescent and the genealogy of continuous state branching processes. *Prob. Theory Relat. Fields* **117**, 249–266 (2000)
24. A. Bovier, I.A.Kurkova, and M. Löwe. Fluctuations of the free energy in the REM and p-spin SK models. *Ann. Probab.* **30**(2), 605–651 (2002)
25. A. Bovier. Statistical mechanics of disordered systems, MaPhySto Lecture Notes 10, Aarhus, (2001). In expanded and updated form: Cambridge Series in Statistical and Probabilistic Mathematics, vol. 18, Cambridge University Press, Cambridge 2006
26. E. Bolthausen and A.-S. Sznitman, *Ten Lectures on Random Media*, Birkhäuser, Basel (2002)
27. M. Talagrand, Mean field models for spin-glasses: a first course. Lectures on probability theory and statistics (Saint-Flour, 2000), 181–285, Lecture Notes in Mathematics, 1816, Springer, Berlin Heidelberg New York, 2003
28. S. Ghirlanda and F. Guerra. General properties of the overlap probability distributions in disordered spin systems. Towards Parisi ultrametricity, *J. Phys. A* **31**, 9144–9155 (1998)
29. M. Talagrand, Rigorous low temperature results for mean field p-spin interaction models, *Prob. Theory. Relat. Fields* **117**, 303–360 (2000)
30. M.R. Leadbetter, G. Lindgren, and H. Rootzén. *Extremes and Related Properties of Random Sequences and Processes*. Springer, Berlin Heidelberg New York (1983)
31. A. Bovier and I. Kurkova. Rigorous results on some simple spin-glass models. *Markov Process. Relat. Fields* **9**(2), 209–242 (2003)
32. M. Ledoux and M. Talagrand, Probability on Banach spaces, Ergebnisse der Mathematik und ihrer Grenzgebiete, vol. 3, Springer, Berlin Heidelberg New York (1991)
33. M.D. Bramson. Maximal displacement of branching Brownian motion, *Commun. Pure Appl. Math.* **31**, 531–581 (1978)
34. J. Pitman. Combinatorial stochastic processes. Lecture Notes of the 2002 Ecole de Probabilités de St. Flour, Springer Lecture Notes in Mathematics

35. J. Pitman, Exchangeable and partially exchangeable random partitions, *Probab. Theory. Relat. Fields* **102**, 145–158 (1995)
36. M. Perman, J. Pitman, and M. Yor, Size-biased sampling of Poisson point processes and excursions, *Probab. Theory. Relat. Fields* **92**, 21–39 (1992)

Dynamics for Spherical Models
of Spin-Glass and Aging

Alice Guionnet

Ecole Normale Supérieure de Lyon, Unité de Mathématiques pures et appliquées
UMR 5669, 46 Allée d'Italie, 69364 Lyon Cedex 07, France
e-mail: aguionne@umpa.ens-lyon.fr

Summary. We review the recent developments in the study of Langevin dynamics
of spin glasses, and in particular for the spherical *p*-spins models of Sherrington–
Kirkpatrick, emphasizing the study of the aging phenomenon. We outline all the
proofs and give a few new arguments.

1 Introduction

The understanding of phase transition is one of the major problem in statistical
physics. It often refers in the literature to the study of the *equilibrium phase
transition*, such as the study of the phase transition of a Gibbs measure (see
for instance Ruelle's book [1]). However, most systems in nature are out of
equilibrium, reason why it is crucial to understand the convergence to equilibrium of dynamics. In the last twenty years, there was a great effort to analyze
this question, in particular by using coercive inequalities such as logarithmic
Sobolev inequalities. In many cases, it could be found that dynamics are well
approximated by equilibrium. However, some physical systems can only be
observed out of equilibrium. Some of them are naturally out of equilibrium
because they are submitted to a gradient of temperature, of potential etc. But
some others, on which we shall focus in this survey, are systems which relax
to equilibrium so slowly that the equilibrium will never be reached during the
experiment or the simulation. Such systems can be encountered in nature;
one of the most well known example is glass (a good reason why glasses refer
in physics to a whole class of systems which never reach equilibrium), but
toothpaste or jelly are other examples. In these cases, it is difficult a priori to
relate the properties of the equilibrium measure and those of the dynamics.
In particular, a notion of *dynamical phase transition* has to be introduced.

Equilibrium phase transition is often characterized by the appearance of
several attracting states for the system, so that the physical medium under
consideration may change dramatically after a very slight modification of
its preparation. A dynamical phase transition could be associated with the

appearance of different time scales along which the evolution of the system would look very different. The notion of aging can serve to characterize the low temperature behaviour of dynamics of such systems; *A system is said to age if the older it gets, the longer it will take to forget its past.* The age of the system is the time spent since the system reached its glass phase, which is often obtained by freezing it below the critical temperature. The experiment exhibiting aging is usually as follows. One considers a medium at high temperature and freeze it at time $t = 0$ at a temperature below the critical temperature T_c. One then measures a parameter $q(t_w, t_w + t)$ where t_w is the age of the system and $t + t_w$ the measurement time. The parameter $q(s, t)$ is often the covariance $E(X_t X_s) - E(X_t)E(X_s)$ of the observable X or the probability $P(X_t = X_s)$. Then, a system is said to age when $q(t_w, t_w + h(t_w))$ converges to a non-zero constant as t_w goes to infinity, for some non-trivial increasing function h going to infinity at infinity. In other words, aging means that there exists non-trivial scales which govern the time correlation of the system. One usually observes the following. At large temperature, the system quickly equilibrates and the order parameter should rapidly become stationary; $q(s, t) \approx q(s - t)$ for t, s reasonably large. At lower temperature, aging will occur if the order parameter q is not a function of $t - t_w$ only, but also depends on the age t_w of the system. On a short scale, when t_w goes to infinity while $\tau = t - t_w$ stays bounded, the dynamics looks stationnary (i.e. only depends on τ) and when then τ goes to infinity reaches a state where q is approximately given by a constant q_{EA}. The system stays in this state quite a long time so that it seems to be in equilibrium. However, on a longer scale, the system will undergo dramatic changes which will drive the parameter q to zero. For instance, in the physics literature, it is expected in many models (such as the spherical pure p-spins models which we shall describe later) that q could "asymptotically" be written under the form, for $s > t$

$$q(s, t) \simeq Q(s - t) + \mathcal{Q}(\frac{h(t)}{h(s)}) \tag{1.1}$$

with some functions $Q(u)$ going to zero as u goes to infinity, $\mathcal{Q}(u)$ going to zero as u goes to zero and some scale function $h(u)$ going to infinity when u does. Such a function satisfies the earlier observations since $\mathcal{Q}(\frac{h(t)}{h(s)})$ is approximately constant (equal to q_{EA}) on the whole time scale $\frac{h(t)}{h(s)} \simeq 1$.

There are many materials that exhibit experimentally a glass phase and aging; let us quote some physics literature on the glass phase of supraconductors [2], granular materials [3, 4], see also the review article [5]. The mathematical literature on the subject is much more scarce; for the time being it is roughly limited to the study of Bouchaud's trap model and its applications on one hand (see [6, 7] for example) and the study of spherical models that we shall describe in this survey on the other hand. Let us also quote the article [10] where a systematic study of aging for certain interaction diffusions was undertaken. In fact, up to now, the study of aging properties have been very dependent on the model, a "slight" generalization of which creating an enormous amount of work to adapt the techniques. Even though the Bouchaud's trap model was designed to mimic the dynamics of the random energy model, the study of these dynamics required infinitely deeper analysis (see [11,12]). Similarly, the 2-spins spherical model of spin glass, that we shall call spherical Sherrington–Kirkpatrick model, is rather easy to study since its covariance satisfies an autonomous equation which can be analysed. As we shall see, its natural generalization to p-spins model (which is the simplest interesting generalization one could think of) results with an autonomous system of integro-differential equations satisfied by the covariance and the so-called response function which is not yet satisfyingly solved, even on a heuristic basis. One of the reason for this complexity is certainly partly due to the fact that (1.1) is too vague; it is understood asymptotically when one consider either the scales where $s - t \ll t$ or $h(t)/h(s) \in (0,1)$ but it does not say anything about the intermediate phase when $s - t \geq t$ and $h(t)/h(s) \simeq 1$. This phase is in fact much more subtle and very rarely described. However, except in very few cases, it should determine the aging function Q (for instance its slope at $u = 1$). This is often referred to as "the matching problem" in physics.

In this survey, we shall consider a very specific model of glass; namely Sherrington–Kirkpatrick model of spin-glass and its spherical versions. Before entering this subject, let us recall the definition of Sherrington–Kirkpatrick model. Hereafter, $N \in \mathbb{N}$ will denote the number of particles. The original Sherrington–Kirkpatrick is described by a quadratic Hamiltonian given by

$$H_{\mathbf{J}}(\mathbf{x}) = \sum_{1 \leq i < j \leq N} J_{ij} x_i x_j$$

where $\mathbf{x} = (x_i, 1 \leq i \leq N)$ represent the particles or spins, which belong to a set M. M can be either discrete, for instance $M = \{-1, +1\}$ in the Ising model, or continuous, for instance $M = \mathbb{R}$ or M is a compact Riemaniann manifold such as a sphere in $\mathbb{R}^d$. The J_{ij}'s are centered independent random variables with variance N^{-1}, often assumed to be Gaussian for simplicity.

It has been generalized into the so-called p-spin model described by the Hamiltonian

$$H_{\mathbf{J}^p}(\mathbf{x}) = \sum_{1 \leq i_1 < i_2 < \cdots < i_p \leq N} J^p_{i_1 \cdots i_p} x_{i_1} x_{i_2} \cdots x_{i_p} \tag{1.2}$$

where $J^p_{i_1 \cdots i_p}$ are centered independent Gaussian variables with covariance N^{-p+1}. In the following, we may also consider mixture of these models given by

$$H_{\mathbf{J}}^{\gamma} = \sum_{p \geq 2} \gamma_p H_{\mathbf{J}^p}$$

where the J^p's are independent from the J^k's if $p \neq k$ and $(\gamma_p)_{p \geq 2}$ is some sequence such that $H_{\mathbf{J}}^{\gamma}$ makes sense.

If μ is a probability measure on M, a Gibbs (or equilibrium) measure for the Sherrington–Kirkpatrick model at temperature $T = \beta^{-1}$ is given by

$$\mu_N(d\mathbf{x}) = \frac{1}{Z_N} e^{-\beta H_J^\gamma(\mathbf{x})} \prod_{i=1}^{N} d\mu(x_i) \text{ with } Z_N = \int e^{-\beta H_{J,N}^\gamma(\mathbf{x})} \prod_{i=1}^{N} d\mu(x_i).$$

The original Sherrington–Kirkpatrick model concerns the case where $\gamma_p = 1_{p=2}$ and the spins are Ising spins, that is $M = \{-1, +1\}$ and μ is the Bernouilli law $\mu(dx) = \frac{1}{2}(\delta_{+1} + \delta_{-1})$. The natural dynamics associated with such dynamics are Glauber dynamics, as considered by Grunwald [13].

In the case $M = \mathbb{R}$ and $\mu(dx) = Z^{-1} e^{-U(x)} dx$ for some smooth potential U, the associated Langevin dynamics (see Sect. 2) were considered by Sompoliski and Zippelius (see [14, 15] and then by Ben Arous and myself [16–19].

In both setting, it is proved that the empirical measure $N^{-1} \sum_{i=1}^{N} \delta_{x_{[0,T]}^i}$ on path-space converges as N goes to infinity for every time $T > 0$ towards some limit law Q_T. Q_T is not Markovian (even though at finite N, the dynamics are Markovian, they lost this property at the limit $N \to \infty$ by self-averaging, the average of Markov laws being not necessarily Markovian) and given by a non-linear equation. This limiting law is so complicated that the behaviour of its covariance could not be analyzed so far, neither in the mathematics or in the physics literature. However, it is believed (see [5]) that the Langevin dynamics for Sherrington–Kirkpatrick dynamics ages and actually with infinitely many time scales. Interestingly, the behaviour of the dynamical covariance is expected to mimic the ultrametricity properties expected [5] for the static; namely it is believed that for large $t_1 < t_3$, one has, with $C(t, s) = \int x_s x_t dQ(x) - \int x_s dQ(x) \int x_t dQ(x)$

$$C(t_1, t_3) \simeq \min_{t_2 \in [t_1, t_3]} \{ C(t_1, t_2), C(t_2, t_3) \}. \tag{1.3}$$

One of the difficulties presented by the analysis of C is that, when the potential U is not quadratic, its evolution depends on another order parameter, namely $C_2(s, t) = \int U'(x_s) x_t dQ(x)$. One can write also an equation for C_2 which will give rise to a third-order parameter, and so forth until one arrives to an infinite system of equations. The idea to simplify the problem is then to change the potential U to get a simpler set of equations; this is what the spherical models are doing as we shall see in Sects. 2 and 3. In fact, the dynamics shall then be represented by only two-order parameters, namely the covariance and the response function. However, when one considers mixture of p-spins (e.g. $\gamma_2 \neq 0$ and $\gamma_4 \neq 0$), the ultrametric picture (1.3) is expected to describe the long time behaviour of the covariance. Hence, interesting dynamical phase transitions are expected to happen also for spherical models. Note that it was already proved for the static in [20] that static phase transition similar to that of the original Sherrington–Kirkpatrick model occurs for spherical models (i.e. with a limiting overlap which takes continuous values, a phenomenon related with continuous symmetry breaking).

In Sects. 2 and 3, I shall describe the spherical Sherrington–Kirkpatrick models and try to describe their long-time dynamics. I hence complement the review article [17] where only Sherrington–Kirkpatrick dynamics and the derivation of the annealed limiting dynamics were described by restricting ourselves to simpler models but emphasizing the analysis of the long times properties of these limiting dynamics. Note here that the question at stake is very different from metastability; the time-scales considered in aging do not depend on the number of particles so that the system cannot visit all its basin of attraction, but rather the neighbourhood of one.

2 Spherical Sherrington–Kirkpatrick Model

If $U : \mathbb{R} \rightarrow \mathbb{R}$ is some potential going to infinity fast enough at infinity, the Langevin dynamics at temperature $T = \beta^{-1}$ for the Sherrington–Kirkpatrick model are defined by the stochastic differential system

$$dx_t^i = -\beta \partial_{x^i} H_J(\mathbf{x}_t)dt - U'(x_t^i)dt + dB_t^i$$

with prescribed initial data. Here, $(B^i, 1 \leq i \leq N)$ are i.i.d Brownian motions. One way to simplify considerably this system is to consider, instead of the single spin constraint $U(x)$, a smooth spherical constraint $U(N^{-1}\sum x_i^2)$. The Langevin dynamics for such a constraint are given by:

$$dx_t^i = -\beta \partial_{x^i} H_J(\mathbf{x}_t)dt - U'(\frac{1}{N}\sum_{j=1}^{N}(x_t^j)^2)x_t^i dt + dB_t^i \qquad (2.1)$$

with a function U on $\mathbb{R}^+$ such that

$$\limsup_{x \rightarrow \infty} \frac{U(x)}{x} = +\infty$$

in order to insure the almost sure boundedness of the empirical covariance under the dynamics (2.1). Note that such dynamics are invariant for the Gibbs measure

$$\mu_N^J(d\mathbf{x}) = Z_{J,N}^{-1}e^{-NU\left(\frac{1}{N}\sum_{j=1}^{N}(x_t^j)^2\right)-2\beta H_J(\mathbf{x})}\prod_{i=1}^{N}dx_i.$$

A hard spherical constraint was considered in [21] where a similar study was undertaken. This hard spherical constraint is not given by considering dynamics on the sphere but rather by replacing $U'(N^{-1}\sum_{j=1}^{N}(x_t^j)^2)$ in (2.1) by a non-random constant $\mu_N(t)$ chosen such that $\mathbb{E}[N^{-1}\sum_{j=1}^{N}(x_t^j)^2] = 1$ at all times t and for all integers N (or at least in the limit N going to infinity). Since what we really care about is long time asymptotics, we see that these two points of view are equivalent provided we can prove that the solution of (2.1) satisfies

$$\lim_{t\to\infty}\lim_{N\to\infty}\frac{1}{N}\sum_{j=1}^{N}(x_t^j)^2 = K \qquad \text{a.s.}$$

for some $K \in \mathbb{R}^{+*}$, and up to the rescaling $x^i \to K^{-\frac{1}{2}}x^i$. The advantage of the smooth spherical constraint (2.1) is that we can prove (see [22]) the existence and uniqueness of the weak solutions of (2.1), that we shall denote $P_{\mathbf{J},\mathbf{x_0}}^N$, as well as boundness properties of the empirical covariance in all L^p and tightness (which we cannot obtain with the model of Cugliandolo and Dean, who assumed from the start convergence of the empirical covariance).

The huge simplification offered by the spherical model is the invariance by the action of the orthogonal group of the potential $U'(N^{-1}\sum_{j=1}^{N}(x_t^j)^2)$; as a consequence, the dynamics of the order parameter given by the empirical covariance

$$K_N(s,t) = \frac{1}{N}\sum_{i=1}^{N}x_s^i x_t^i \tag{2.2}$$

depend mostly on the eigenvalues of the random interacting potential $\mathbf{J}$ and satisfies, at the large N limit, an autonomous equation. In [22], this convergence was obtained by means of a large deviation principle. I here give a more direct proof of this convergence, inspired by the proof followed in [23] for the p-spins model. It is not based on a cavity method as developed in physics, but it follows the standard strategy to prove tightness of the order parameters and then uniqueness of the limit points due to their characterization as the unique solution to a certain integro-differential equation.

Theorem 2.1. *Assume that $(x_0^i, 1 \le i \le N)$ are independent equidistributed variables with law μ_0 so that*

$$\int e^{\alpha x^2}d\mu_0(x) < \infty$$

for some $\alpha > 0$. For any time $T \in \mathbb{R}^+$, $K_N \in \mathcal{C}([0,T]\times[0,T],\mathbb{R})$ converges almost surely towards K, solution of the renewal equation

$$K(t,s) = R(t)^{-\frac{1}{2}}R(s)^{-\frac{1}{2}}\mathcal{L}(\beta(t+s))\int x^2 d\mu_0(x)$$

$$+ \int_0^s \frac{R(v)}{\sqrt{R(t)R(s)}}\mathcal{L}(\beta(t+s-2v))dv \tag{2.3}$$

with, for $\theta > 0$ and $t \geq 0$,

$$\mathcal{L}(\theta) = \int e^{\theta\lambda} d\sigma(\lambda)$$

$$R(t) = e^{2\int_0^t U'(K(s))ds}$$

where $K(s) = K(s,s)$ for $s \geq 0$.

Proof. The proof is rather classical; one first shows that $(K_N(u) = K_N(u,u), u \leq T)$ is integrable and self-averages around its mean. One then proves that its mean is tight and actually converges. The convergence of $(K_N(u), u \leq T)$ then results with the convergence of $(K_N(u,t), u,t \leq T)$:

1. *Almost sure tightness* Recall that by Arzela–Ascoli's theorem, compact sets of $\mathcal{C}([0,T] \times [0,T], \mathbb{R})$ are of the form:

$$K_{M,(\varepsilon,\delta)} = \Big\{ f : [0,T] \times [0,T] \to \mathbb{R}, \sup_{s \leq T} |f(s)|$$

$$\leq M, \sup_{|s-t| \leq \varepsilon(\delta)} |f(s) - f(t)| \leq \delta, \forall \delta > 0 \Big\}$$

where M is some finite constant, $\epsilon(\delta) > 0$ and ε can be chosen arbitrary small. To show that K_N belongs to such a set with high probability, observe first that (2.1) implies by Itô's formula that

$$K_N(t) = K_N(0) - 2\int_0^t [U'(K_N(u))K_N(u)$$

$$+ \frac{\beta}{N} \sum_{i=1}^N J_{ij} x_u^i x_u^j] du + t + \frac{2}{N} \sum_{i=1}^N \int_0^t x_u^i dB_u^i. \qquad (2.4)$$

Now, by Doob's inequality applied to the martingale $L_t^N = \exp\{\sum_{i=1}^N [\int_0^t x_u^i dB_u^i - \frac{1}{2}\int_0^t (x_u^i)^2 du]\}$, we get that for all $z \geq 0$, if $\Omega_{N,z} = \{x : \sup_{t \leq T} \{\frac{1}{N}\sum_{i=1}^N \int_0^t x_u^i dB_u^i - \frac{1}{2N}\int_0^t K_N(u)du\} \leq z\}$,

$$P(\Omega_{N,z}^c) \leq e^{-Nz}.$$

Remark that U' is bounded below by a constant U by assumption and denote by $\|\mathbf{J}\|_\infty$ the spectral radius of $\mathbf{J}$. Then, on $\Omega_{N,z}$, we deduce from (2.4) that

$$K_N(t) \leq K_N(0) + 2\int_0^t [\beta\|\mathbf{J}\|_\infty - U + \frac{1}{N}]K_N(u)du + t + 2zt.$$

Hence, by Gronwall's lemma, on $\Omega_{N,z}$,

$$\sup_{t \le T} K_N(t) \le (K_N(0) + T + 2zT)e^{(\beta \|\mathbf{J}\|_\infty - U)T}$$

and so there exists $a = a(T, U) > 0$ so that for any x

$$P\left(\sup_{t \le T} K_N(t) \ge x\right) \le P(\Omega^c_{N,ax}) + P(K_N(0) \ge ax) + P(\|\mathbf{J}\|_\infty \ge a \log x)$$

$$(2.5)$$

By Chebychev's inequality, we conclude that

$$P(K_N(0) \ge ax) \le \left(\int e^{\alpha x^2} d\mu_0(x)\right)^N e^{-a\alpha xN}.$$

Moreover, it is easy to see that

$$\|\mathbf{J}\|_\infty = \sup_{\|u\|=1} < u, Ju >$$

has sub-Gaussian tail. Indeed, it is a Lipschitz function of the Gaussian entries of $\mathbf{J}$ (with respect to the Euclidean norm) with constant bounded by $N^{-\frac{1}{2}}$ so that by Herbst argument (see [24,25]), we have the concentration inequality

$$P(\|\mathbf{J}\|_\infty \ge E[\|\mathbf{J}\|_\infty] + x) \le e^{-\frac{1}{2}Nx^2}.$$

Thus, since it is well known that $E[\|\mathbf{J}\|_\infty]$ converges towards 2 (see e.g. [26] and [27]), we deduce that for large x

$$P(\|\mathbf{J}\|_\infty \ge a \log x) \le e^{-\frac{N}{3}(a \log x)^2}.$$

Hence, we conclude from (2.5) that for large x,

$$P\left(\sup_{t \le T} K_N(t) \ge x\right) \le e^{-\frac{N}{4}(a \log x)^2}$$

$$(2.6)$$

In particular, $\sup_{t \le T} K_N(t)$ is integrable.

Similarly, for $s \ge t$,

$$K_N(s, t) - K(t, t) = \int_t^s [-U'(K_N(u))K_N(u, t)$$

$$+ < \mathbf{J}x_u, x_t >]du + \frac{1}{N} \sum_{i=1}^N x_t^i (B_s^i - B_t^i).$$

Now, it is well known (see [28]) that for some $\alpha > 0$

$$e^L = P\left(e^{\alpha \sup_{|s-t| \le \delta} \frac{|B_s^i - B_t^i|^2}{s-t}}\right) < \infty$$

so that

$$P(\sup_{|s-t|\leq\delta} |\frac{1}{N}\sum_{i=1}^{N} x_t^i(B_s^i - B_t^i)| \geq \varepsilon)$$

$$\leq P(\sup_{|s-t|\leq\delta} \frac{1}{N}\sum_{i=1}^{N}(B_s^i - B_t^i)^2 \geq \frac{\varepsilon^2}{M}) + P(\sup_{t\leq T} K_N(t) \geq M)$$

$$\leq e^{-\alpha\frac{\varepsilon^2}{M\delta}N} E[e^{\alpha\sum_{i=1}^{N}\sup_{|s-t|\leq\delta}\frac{|B_s^i - B_t^i|^2}{s-t}}] + e^{-cN(\log M)^2}$$

$$\leq e^{-\alpha\frac{\varepsilon^2}{M\delta}N+LN} + e^{-cN(\log M)^2}.$$

Hence, if $\delta_n = n^{-1}$, $\varepsilon_n = n^{-\frac{1}{4}}$, $M = n^{\frac{1}{4}}$ we get for N large enough

$$P(\sup_{|s-t|\leq\delta_n} |\frac{1}{N}\sum_{i=1}^{N} x_t^i(B_s^i - B_t^i)| \geq \varepsilon_n) \leq e^{-c'(\log n)^2 N}$$

and so by Borel–Cantelli Lemma

$$P(\cup_{N_0} \cap_{N\geq N_0} \{K_N \in K_{M,(\varepsilon,\delta)}\}) = 1$$

which proves that K_N is almost surely tight. Its mean is as well tight according to the earlier tails estimates.

2. *Self-averaging of* $(K_N(u), u \leq T)$. Going back to (2.1), we find that

$$x_t = e^{-\int_0^t \mathbf{V_N}(u)du} x_0 + \int_0^t e^{-\int_v^t \mathbf{V_N}(u)du} dB_v$$

with $\mathbf{V_N}(u)$ the $N \times N$ matrix given by $\mathbf{V_N}(u) := U'(K_N(u,u))\mathbf{I} - \beta\mathbf{J}$ if $\mathbf{J}$ is the symmetric matrix with entries $\{J_{ij}, 1 \leq i \leq j \leq N\}$ above the diagonal. Note here that the $\{\mathbf{V_N}(u), u \geq 0\}$ commute so that the exponential is taken in the usual sense and that

$$\int_0^t e^{-\int_v^t \mathbf{V_N}(u)du} dB_v = e^{-\int_0^t \mathbf{V_N}(u)du} \int_0^t e^{\int_0^v \mathbf{V_N}(u)du} dB_v$$

is well defined eventhough $\int_v^t \mathbf{V_N}(u)du$ is not adapted. Therefore,

$$K_N(t) = \frac{1}{N}||e^{-\int_0^t \mathbf{V_N}(u)du} x_0 + \int_0^t e^{-\int_v^t \mathbf{V_N}(u)du} dB_v||^2.$$

Let $k_N(t)$ be obtained by replacing in $K_N(t)$, K_N by its expectation in $\mathbf{V_N}(u)$. Namely for $0 \leq t \leq T$, we set

$$k_N(t) = \frac{1}{N}||e^{-\int_0^t (U'(\mathbb{E}[k_N(u)])I-\beta\mathbf{J})du} x_0 + \int_0^t e^{-\int_v^t (U'(\mathbb{E}[k_N(u)])I-\beta\mathbf{J})du} dB_v||^2.$$

$$(2.7)$$

Note that

$$\mathbb{E}[k_N(t)] = \mathbb{E}\Big[\frac{1}{N} < x_0, e^{-2\int_0^t (U'(\mathbb{E}[k_N(u)])I - \beta \mathbf{J})du} x_0 >$$
$$+ \int_0^t \frac{1}{N}\mathrm{tr}\big(e^{-2\int_v^t (U'(\mathbb{E}[k_N(u)])I - \beta \mathbf{J})du}\big)dv\Big]$$

is uniformly bounded on $t \leq T$ (since $e^{||\mathbf{J}||_\infty}$ has finite moments and U' is bounded below). It is easy to see, by a standard fixed point argument, that $\mathbb{E}[k_N(t)]$ defined by the earlier equation exists and is unique, from which existence of $(k_N(t), t \geq 0)$ is clear. Moreover, since $(\mathbf{J}, B, x_0)$ are all independent, $(k_N(t), t \leq T)$ will self-average. In fact, using Doob's inequality again, we can prove that for all $T < \infty$, there exists

$$\mathbb{E}\Big[\sup_{t \leq T}(k_N(t) - \mathbb{E}[k_N(t)])^4\Big] \leq \frac{C(T)}{N^2}.$$

Moreover, getting rid of the stochastic integral by integration by parts, we see that on the set $\{\sup_{t \leq T} K_N(t) \leq M\} \cap \{||\mathbf{J}||_\infty \leq M\} \cap \{N^{-1}||x_0||^2 \leq M\} \cap \{\sup_{t \leq T} N^{-1}||\mathbf{B}_t||^2 \leq M\}$ with overwhelming probability,

$$|K_N(t) - k_N(t)| \leq C(M,T) \int_0^t |K_N(t) - \mathbb{E}[k_N(t)]|dt.$$

Thus, since $\sup_{t \leq T} |\mathbb{E}[k_N(t)] - k_N(t)|$ goes to zero almost surely, we conclude by Gronwall's lemma that

$$\lim_{N \to \infty} \sup_{t \leq T} |K_N(t) - k_N(t)| = \lim_{N \to \infty} \sup_{t \leq T} |K_N(t) - \mathbb{E}[k_N(t)]| = 0 \text{ a.s..}$$

Since $(K_N(t), t \leq T)$ is uniformly integrable, we conclude also that $E[K_N(t)] - \mathbb{E}[k_N(t)]$ converges towards zero uniformly on $t \leq T$.

3. *Equations for the limit points.* Since $(\mathbb{E}[K_N(t)], t \leq T)$ is tight, so does $(\mathbb{E}[k_N(t)], t \leq T)$. Let K be a limit point. To obtain an equation for K from (2.7), note that by independence of $\mathbf{J}$ and $\mathbf{x}_0$, if $(\lambda_i)_{1 \leq i \leq N}$ denotes the eigenvalues of $\mathbf{J}$,

$$\mathrm{Law}\Big(\frac{1}{N} < x_0, e^{2t\beta \mathbf{J}} x_0 >\Big) = \mathrm{Law}\Big(\frac{1}{N} < x_0, x_0 > \sum_{i=1}^N e^{2t\beta \lambda_i} u_i^2\Big)$$

where u is independent of the eigenvalues λ of $\mathbf{J}$ and of x_0 and follows the uniform law on the sphere $S_{\sqrt{N}}^{N-1}$ with radius $\sqrt{N}$. Since $(u_i^2)_{1 \leq i \leq N}$ has the same law that $((\sum_{j=1}^N g_j^2)^{-1} g_i^2)_{1 \leq i \leq N}$ for independent Gaussian random variables g, the asymptotics are derived by standard law of large numbers and we find that as $N^{-1} \sum \delta_{\lambda_i}$ converges towards $\sigma(dx) = C\sqrt{4 - x^2}dx$ according to Wigner [29], when $(\mathbb{E}[K_N'(u)], u \leq T)$ converges to $(K(t), t \leq T)$,

$$\lim_{N\to\infty} \frac{1}{N} ||e^{-\int_0^t (U'(\mathbb{E}[K'_N(u)])I-\beta\mathbf{J})du} x_0||^2 = e^{-2\int_0^t U'(K(u))du} \mathcal{L}(2\beta t) \int x^2 d\mu_0(x)$$

Hence, by (3.6), we get

$$K(t) = e^{-2\int_0^t U'(K(u))du} \mathcal{L}(2\beta t) \int x^2 d\mu_0(x)$$
$$+ \int_0^t e^{-2\int_v^t U'(K(u))du} \mathcal{L}(2\beta(v-t))dv. \qquad (2.8)$$

Now, we know that all limit points are uniformly bounded by the first subsection (at least on compact times intervals) and it is easy to see that (2.8) has a unique solution which is uniformly bounded by a given constant on compact sets (by a standard fixed point argument). This gives the convergence of K_N almost surely and in expectation in $\mathcal{C}([0,T],\mathbb{R})$ towards K, the unique solution of (2.8).

4. *Convergence of* $K_N \in \mathcal{C}([0,T],[0,T],\mathbb{R})$. Now, again from (2.1)

$$K_N(t,s) = \frac{1}{N}\mathrm{tr}\left(\left(e^{-\int_0^t \mathbf{V_N}(u)du} x_0 + \int_0^t e^{-\int_v^t \mathbf{V_N}(u)du} dB_v\right)\right.$$
$$\left. \times \left(e^{-\int_0^s \mathbf{V_N}(u)du} x_0 + \int_0^s e^{-\int_v^s \mathbf{V_N}(u)du} dB_v\right)\right)$$

and, as earlier, we can replace, in $\mathbf{V_N}(u)$, K_N by its limit so that

$$K_N(t,s) \simeq \frac{1}{N}\mathrm{tr}\left(\left(e^{-\int_0^t (U'(K(u))I-\beta\mathbf{J})du} x_0 + \int_0^t e^{-\int_v^t (U'(K(u))I-\beta\mathbf{J})du} dB_v\right)\right.$$
$$\left. \times \left(e^{-\int_0^s (U'(K(u))I-\beta\mathbf{J})du} x_0 + \int_0^s e^{-\int_v^s (U'(K(u))I-\beta\mathbf{J})du} dB_v\right)\right)$$
$$+ o(N^{-1}).$$

It is then easy to obtain the convergence of $K_N(t,s)$ by usual law of large numbers towards

$$K(t,s) = e^{-\int_0^t U'(K(u))du - \int_0^s U'(K(u))du} \mathcal{L}(\beta(t+s)) \int x^2 d\mu_0(x)$$
$$+ \int_0^t e^{-\int_v^t (U'(K(u)))du - \int_v^s U'(K(u)))dudu} \mathcal{L}(\beta(2v-t-s))dv.$$

Since K_N and $E[K_N]$ are tight in $\mathcal{C}([0,T]\times[0,T],\mathbb{R})$ we can conclude that K_N and $E[K_N]$ converge towards K in $\mathcal{C}([0,T]\times[0,T],\mathbb{R})$ which finishes the proof.

Note that similar convergence results can be obtained when one starts from different initial conditions provided that

$$< \mathbf{x}_0, e^{\theta \mathbf{J}} \mathbf{x}_0 > \qquad (2.9)$$

converges for all $\theta \in \mathbb{R}$.

For instance, if $\mathbf{J} = O^* DO$ with a diagonal matrix D with eigenvalues $(d_i)_{1 \leq i \leq N}$ so that $N^{-1} \sum_{i=1}^N d_i$ converges towards a probability measure σ and O follows the Haar measure on the orthogonal group, we obtain exactly the same type of results but with $\mathcal{L}$ the Laplace transform of the measure σ (we just need that the spectral radius does not grow too fast to insure tightness of K_N and convergence of the Laplace transform $N^{-1} \mathrm{tr}(e^{\theta \mathbf{J}})$ (this actually improves on the hypotheses of [22] where we needed to assume additionally that the spectral radius of $\mathbf{J}$ converges towards the edge of σ to use large deviations techniques)).

If the initial conditions is taken to be an eigenvector of $\mathbf{J}$ for an eigenvalue λ_i^N of $\mathbf{J}$ converging towards $x \in [-2, 2]$, we find that K_N converges towards the unique solution of

$$K(t, s) = e^{- \int_0^t U'(K(u)) du - \int_0^s U'(K(u)) du} e^{-\beta(t+s)x}$$
$$+ \int_0^t e^{- \int_v^t U'(K(u)) du - \int_v^s U'(K(u))) du} \mathcal{L}(\beta(2v - t - s)) dv. \quad (2.10)$$

If $\mathbf{x}_0$ follows the Gibbs measure $\mu_N^{\mathbf{J}}$, we can also obtain the convergence of (2.9) by large deviation techniques (see [22]) and hence the convergence of the empirical covariance towards $K_{\mathrm{stat}}(t, s) = K_{\mathrm{stat}}(|t - s|)$. Note that $\mu_N^{\mathbf{J}}$ can be precisely analyzed in this particular model (see [22]).

The interesting point in this model is that we can exactly study the large time behaviour of $K(s, t)$ when t and s go to infinity. In view of (2.3), it is governed by the asymptotics of the Laplace transform $\mathcal{L}$ of σ, which is given, as well as the asymptotic behaviour of $K(u)$ as u goes to infinity. The difficulty here is that we not only need to find the first-order term in $K(u)$ as u goes to infinity but also its correction at least in the low temperature phase where β is large (since we expect $K(t, s)$ to exhibit a polynomial decay then). We consider later general compactly supported probability measure σ which is symmetric, $\sigma(x \in .) = \sigma(-x \in .)$ (the last assumption to lighten notations). We proved in [22] the following: Assume

$$U(x) = \frac{c}{2} x^2$$

for some $c > 0$. Let λ^* be the smallest real number so that $\sigma((\lambda^*, \infty)) = 0$ ($\lambda^* = 2$ when σ is the semi-circular law). Then,

Theorem 2.2. *Let*

$$\beta_c^2 = \frac{c}{2\lambda^*} \int \frac{1}{\lambda^* - \lambda} d\sigma(\lambda). \qquad (2.11)$$

and assume that

$$\sigma(\lambda^* - \lambda \leq \theta) \simeq_{\theta \to 0} b_1 \theta^q \qquad (2.12)$$

for some $q > 1$ and $b_1 > 0$ (and hence $\beta_c < \infty$). If σ is the semi-circular law, (2.12) is satisfied with $q = \frac{3}{2}$. Then,

(A) When the initial data follows the Gibbs measure, the limiting covariance is stationnary and given by K_{stat} which as the explicit form

$$K_{\text{stat}}(\tau) = q_{EA}(\beta) + C \int_{[-2,2]} (s_\beta - \lambda)^{-1} e^{-(s_\beta - \lambda)\tau} \sqrt{4 - \lambda^2} d\lambda$$

with a finite constant C, $s_\beta = 2$ for $\beta > \beta_c$ and $s_\beta > 2$ when $\beta < \beta_c$, $q_{EA}(\beta) = 0$ for $\beta < \beta_c$ and $q_{EA}(\beta) > 0$ when $\beta > \beta_c$.

(B) When starting from i.i.d initial conditions, the unique solution K to (2.3) satisfies

1. For $\beta < \beta_c$, there exists $\delta_\beta > 0$ and $c_\beta \in \mathbb{R}^+$ so that for all $t, s \in \mathbb{R}^+$,

$$|K(t,s)| \leq c_\beta e^{-\delta_\beta |t-s|}. \qquad (2.13)$$

2. For $\beta = \beta_c$, $q \neq 2$, $t \gg s \gg 1$, we have the polynomial decay

$$K(t,s) \sim \begin{cases} (t-s)^{1-q} & \text{for } t/s \text{ bounded} \\ \dfrac{s^{1-\psi_q/2}}{t^{q-\psi_q/2}} & \text{otherwise}, \end{cases} \qquad (2.14)$$

   where $\psi_q = \max(2 - q, 0)$.

3. When $\beta > \beta_c$ we get that

   a) for all $\tau \in \mathbb{R}^+$,

$$\lim_{t \to \infty} K(t, t+\tau) = K_{\text{stat}}(\tau).$$

   In particular,

$$\lim_{\tau \to \infty} \lim_{t \to \infty} K(t, t+\tau) = q_{EA}(\beta) \neq 0.$$

   b) for all $\theta \in [0,1]$,

$$\lim_{\substack{s,t \to \infty \\ s/t \to \theta}} K(t,s) = C\theta^{\frac{q}{2}} \qquad (2.15)$$

   for a positive constant C.

   so $K(t,s) \to 0$ if and only if $t/s \to \infty$.

For this simple model, the static phase transition (see [22]) happens exactly at the same value β_c. In general, one expects the dynamical phase transition to happen before the static phase transition. For β smaller than the critical value for the dynamics, it is expected that $K(t, t+\tau)$ converges as t goes to infinity towards the stationary covariance. However, one should not expect this to be true below the critical temperature in general (on the contrary to what happens here, see (B).3.(a)).

The idea of the proof is to remark that the asymptotics of $K(t, s)$ are governed by the precise asymptotics of

$$R(t) = e^{2 \int_0^t U'(K(u))du}$$

as t goes to infinity, that is by the asymptotic behaviour of K on the diagonal, up to its second-order correction. When $U(x) = \frac{c}{2}x^2$,

$$K(t)e^{2 \int_0^t U'(K(u))du} = K(t)e^{2c \int_0^t K(u)du} = \frac{1}{2c}\partial_t e^{2c \int_0^t K(u)du}$$

so that the equation for K becomes, in terms of $R(t) = e^{2 \int_0^t U'(K(u))du}$, the linear Volterra integrodifferential equation,

$$R'(t) = 2cK(t)R(t) = 2c\mathcal{L}(2\beta t) + 2c \int_0^t R(\tau)\mathcal{L}(2\beta(t - \tau))d\tau . \qquad (2.16)$$

The asymptotic behaviour of solutions to such equations can then be studied by taking the Laplace's transform and using Tauberain's theorem ideology. Let us define s_β to be the unique solution of

$$p(s, \beta) = \frac{2\beta^2}{c}s - \int \frac{1}{s - \lambda}d\sigma(\lambda) = 0$$

when $\beta < \beta_c$, whereas $s_\beta = \lambda^*$ for $\beta > \beta_c$. Set

$$g(\tau) = e^{-2\beta s_\beta \tau} R(\tau) \text{ and } \boldsymbol{g}(z) := \int_0^\infty e^{-2\beta z \tau} g(\tau)d\tau$$

for z with strictly positive real part. Then, one deduces from (2.16) that, if $\mathbf{L}(s) := \int \frac{1}{s-\lambda}d\sigma(\lambda)$,

$$\boldsymbol{g}(z) = \frac{(\mu_0(x^2)c\mathbf{L}(s_\beta + z) + \beta)}{cp(s_\beta + z, \beta)} . \qquad (2.17)$$

Note that for $\beta < \beta_c$, $s_\beta > \lambda^*$ (and so s_β is a simple pole of $p(., \beta)$) whereas $s_\beta = \lambda^*$ for $\beta > \beta_c$ and then because of the tail of σ at λ^* (which implies that $\mathbf{L}^{([q])}(z + \lambda^*) \simeq az^{q-[q]+1}$),

$$z^{[q]+1-q}\boldsymbol{g}^{([q])}(z) \simeq \xi ,$$

where $\xi = b_2(\beta + c\boldsymbol{g}(0))/(cp(\lambda^*, \beta)) \neq 0$. Using complex analysis (note here that standard Tauberian Theorems are not sufficient to get such precise estimates) one then deduces (see [22], Lemma 7.2) that

$$R(x) \simeq_{x \uparrow \infty} C_{q,\beta} x^{-\psi} e^{2s_\beta x} . \tag{2.18}$$

with $\psi = 0$ for $\beta < \beta_c$, $\psi = \psi_q$ for $\beta = \beta_c$, $q \neq 2$, and $\psi = q$ for $\beta > \beta_c$ and some $C_{q,\beta} \in (0, \infty)$. Plugging this estimate into (2.3) yields the lemma.   □

Remarks

1. Similarly, if the initial condition is taken to be an eigenvector of $\mathbf{J}$ with eigenvalue converging to λ^*, we can analyze the solution of (2.10) and see that the aging phenomenon does not appear. That is, $K(t, s) \rightarrow q_{EA} \in (0, \infty)$ for any $\beta > \beta_c$, regardless of how $t - s$ and s approach infinity. The Edwards–Anderson parameter q_{EA} is also the limit of $K_{\text{stat}}(|t - s|)$ when $|t - s|$ goes to infinity, when starting from the invariant measure μ_N^J.
2. The condition $U(x) = 2^{-1}cx^2$ could a priori be relaxed. The main difficulty is then to show the convergence of $K(t)$ as t goes to infinity. But once one supposes that $K(t)$ converges to $K(\infty)$ as t goes to infinity, it is possible, when for instance U is strictly convex (and so U' is inversible) and three times continuously differentiable, to make the following heuristics. Consider the expansion

$$K(t) = A + cU'(K(t)) + O((U'(K(t)) - U'(K(\infty)))^2).$$

Here, $A = K(\infty) - U'(K(\infty))/U''(K(\infty))$ and $c = U''(K(\infty))^{-1}$. One then arrives at the approximate equation

$$A + cU'(K(t)) \approx e^{-2 \int_0^t U'(K(u))du} \mathcal{L}(2\beta t) \int x^2 d\mu_0(x)$$
$$+ \int_0^t e^{-2 \int_v^t (U'(K(u)))I)du} \mathcal{L}(2\beta(v - t))dv$$

and so with $g(t) = e^{2 \int_0^t U'(K(u))du}$, and $\mathbf{g}$ its Fourier transform as earlier, we deduce from (2.16) that

$$\mathbf{g}(z) \simeq \frac{(c\mathbf{L}(z) + \beta)}{c\bar{p}(z, \beta)} \tag{2.19}$$

with $\bar{p}(z, \beta) = c^{-1}A + 2c^{-1}\beta z - \mathbf{L}(z)$. Here, since the equivalence is taken up to the second-order in $K(t)$, $\mathbf{g}(z)$ is given up to its second-order in z. From here, the arguments of the previous proof can be extended up to the first-order correction.
3. The order parameter

$$\bar{K}(t, s) = \lim_{N \to \infty} \left(K_N(t, s) - \left(\frac{1}{N} \sum_{i=1}^N x_t^i\right)\left(\frac{1}{N} \sum_{i=1}^N x_s^i\right) \right)$$

is a priori more natural to consider than K since it clearly measures the long-time correlations. However, it was shown in [22], Sect. 3.6, that $\bar{K}$

is equal to K when the initial conditions are either given by the Gibbs measure or by an eigenvector of $\mathbf{J}$, and in the case of i.i.d initial conditions

$$\bar{K}(t,s) = K(t,s) - \mu_0(x^2)\frac{\mathcal{L}(t)\mathcal{L}(s)}{\sqrt{R(t)}\sqrt{R(s)}}.$$

Even in this case, we deduce from the asymptotics of $R(t)$ that the long time behaviour of $\bar{K}(t,s)$ is governed by that of $K(t,s)$.

3 Spherical p-spins Model

The spherical p spins-model is described by the Hamiltonian $H_{\mathbf{J}^p}$ described in (1.2) with a spherical constraint. The Gibbs measure for a mixture of p-spins Sherrington–Kirkpatrick models can be defined by

$$\mu_N^{\mathbf{J},\gamma}(d\mathbf{x}) = \frac{1}{Z_N^{\mathbf{J},\gamma}} \exp\left\{-\beta H_{\mathbf{J}}^\gamma(\mathbf{x}) - NU\left(\frac{1}{N}\|\mathbf{x}\|^2\right)\right\}d\mathbf{x}$$

for some sequence γ such that $\gamma_p = 0$ for $p \geq M$ for some finite M and a potential U going to infinity fast enough to insure that $\mu_N^{\mathbf{J},\gamma}$ is well defined. To be more precise, we will assume later that the variance of $J_{i_1\ldots i_p}$ is $c(\{i_1,\ldots,i_p\})N^{-p+1}$, where

$$c(\{i_1,\ldots,i_p\}) = \prod_k l_k!, \tag{3.1}$$

and $(l_1, l_2, \ldots)$ are the multiplicities of the different elements of the set $\{i_1, \ldots, i_p\}$ (for example, $c = 1$ when $i_j \neq i_{j'}$ for any $j \neq j'$, while $c = p!$ when all i_j values are the same).

Talagrand considered recently in [20] such a Gibbs measure but with a hard spherical constraint $U(x) = +\infty$ if $x \neq 1$ (more precisely by taking $\mathbf{x}$ uniformly distributed on $S_{\sqrt{N}}^{N-1}$). He obtained, at all temperature $T = \beta^{-1} > 0$, the convergence of the free energy as well as a description of the limiting order parameter

$$q_{1,2} = \lim_{N\to\infty} \frac{1}{N}\sum_{i=1}^N x_i^1 x_i^2$$

where $\mathbf{x}^1$ and $\mathbf{x}^2$ are two independent copies of $\mathbf{x}$ with law $\mu_N^{\mathbf{J},\gamma}$ (but with the same random coupling $\mathbf{J}$). For technical reasons, he proved this convergence under the assumption that all γ_i are not zero. The limiting free energy depends on the model via the function

$$\nu(x) = \sum_{p\geq 1} \frac{\gamma_p^2}{p!} x^p$$

and the temperature (which could actually be incorporated in the $\gamma's$). At low temperature (i.e. large β) the law of $q_{1,2}$ becomes non-trivial and depends a lot on properties of the function ν. Following Crisanti and Sommers, Talagrand proved in Proposition 2.2 of [20] that if $\phi(x) = \nu''(x)^{-\frac{1}{2}}$ is convex, then the law of q_{12} is at most concentrated in two points. This is in particular the case when one considers a pure p-spins where $\nu(x) = x^p$. However, for some ν, the law Q of q_{12} can charge a whole interval; following Talagrand [20], if one takes $\nu(x) = \beta^2(\mathrm{ch}(x) - 1)$, for $\beta > 1$, then for q_0 the unique root of $\phi(q_0) = 1 - q_0$ and $\phi'(q_0) \geq -1$, $Q([0, x]) = -\phi'(x)$ for $x \leq q_0$. Playing with the function ν, it is believed that spherical models can exhibit very sophisticated phase transitions (for instance, one expects appropriate mixture of 2- and 4-spins model to exhibit the same type of phase transitions than the original Sherrington–Kirkpatrick model).

Cugliandolo and Kurchan undertook the study of the corresponding dynamics in [30, 31] with again dynamics of the form

$$dx_t = dB_t - \mu_N(t)x_t dt + \beta \nabla H_J^\gamma(x_t)dt$$

with $\mu_N(t)$ a parameter chosen so that $E[||x_t||^2] = N$ for all $t > 0$. They derived a set of equations for the limiting empirical covariance and the so-called response function. The response function is defined as follows. Let $P_{J,h}^N$ be the weak solution of

$$dx_t = dB_t - \mu_N(t)x_t dt + \beta \nabla H_J^\gamma(x_t)dt + h_t dt$$

for a bounded adapted field h. Then, the response function for the N-particles system is given, for $s < t$, by

$$R_N(t, s) = \frac{1}{N} \sum_{i=1}^{N} \partial_{h_s^i} \int x_t^i dP_{J,h}^N(x)|_{h=0},$$

that is R_N is the mean response of the system to a field perturbation. This definition calls for a heuristic explanation. One can discuss the existence of $R_N(t, s)$, and find an alternative definition for R_N, by using Girsanov's formula. Indeed, it shows that for any adapted bounded process h with values in $\mathbb{R}^N$

$$\int x_t^i dP_{J,h}^N(x) = \int x_t^i e^{\int_0^t h_u . dB_u - \frac{1}{2}\int_0^t ||h_s||^2 ds} dP_{J,0}^N(x)$$

$$\approx \int x_t^i dP_{J,0}^N(x) + \int x_t^i \int_0^t h_u . dB_u dP_{J,0}^N(x) + O(||h||^2).$$

Thus, if $h = h^i$ with $h_j^i = 0$ for $j \neq i$ and $h_i^i = g$ a bounded deterministic function, we find that $\varepsilon \to \int x_t^i dP_{J,\varepsilon h^i}^N(x)$ is differentiable at the origin and with derivative

$$\partial_\varepsilon \int x_t^i dP_{\mathbf{J},\varepsilon h^i}^N(\mathbf{x})|_{\varepsilon=0} = \int x_t^i \int_0^t g_u dB_u^i dP_{\mathbf{J},0}^N(\mathbf{x}).$$

This justifies that $g_s \to \int x_t^i dP_{\mathbf{J},h^i}^N(\mathbf{x})$ is differentiable, with derivative given in the sense of distributions by

$$R_N^i(t,s) := \partial_s \int \frac{1}{N} \sum_{i=1}^N x_t^i B_s^i dP_{\mathbf{J},0}^N(\mathbf{x}).$$

Similarly, we find that $\psi_t : g \to \frac{1}{N} \sum_{i=1}^N \int x_t^i dP_{\mathbf{J},h^i}^N(\mathbf{x})$ is differentiable in the sense of distribution with

$$\int_0^t R_N(t,s)g_s ds := \int_0^t \frac{\partial \psi_t}{\partial g_s}|_{g=0} g_s ds = \frac{1}{N} \sum_{i=1}^N \int x_t^i \int_0^t h_u dB_u^i dP_{\mathbf{J},0}^N(\mathbf{x}),$$

and so we have justified that, in the sense of distributions,

$$R_N(t,s) = \partial_s \int \frac{1}{N} \sum_{i=1}^N x_t^i B_s^i dP_{\mathbf{J},0}^N(\mathbf{x}). \qquad (3.2)$$

In the following, we shall not use the interpretation of R_N as a response function but rather the earlier definition. According to Cugliandolo and Kurchan, $R_N(t,s)$ and $K_N(t,s)$ converge almost surely for all times t,s and the limits R,K satisfies a system of integro-differential equations. In [23], the same problem was considered but with a smooth spherical constraint, that is we considered the stochastic differential system

$$d\mathbf{x}_t = d\mathbf{B}_t - U'(N^{-1}||\mathbf{x}_t||^2)\mathbf{x}_t dt + \nabla H_{\mathbf{J}}^\gamma(\mathbf{x}_t)dt \qquad (3.3)$$

where for some $m \in \mathbb{N}$, $\gamma_j = 0$ for $j > m$ and U satisfies

$$\inf_{\rho \geq 0}\{U'(\rho) - A\rho^{m/2+\delta-1}\} > -\infty. \qquad (3.4)$$

Note here that the Gibbs measure considered by Talagrand with the hard-sphere constraint is not reversible for the dynamics considered by Cugliandolo and Kurchan; for that, one should rather consider dynamics on the sphere. The dynamics considered in [23] and described by (3.3) are reversible with respect to $\mu_N^{\mathbf{J},\gamma}$.

In [23], we could prove the results of Cugliandolo and Kurchan for the system (3.3). Namely, we assumed that the initial condition $\mathbf{x}_0$ is independent of the disorder $\mathbf{J}$, with i.i.d entries with law μ_0. We additionally suppose here that μ_0 satisfies a log-Sobolev inequality, to insure good concentration properties (see [23] for weaker hypotheses). We let χ_N be the integrated response function

$$\chi_N(t,s) = \frac{1}{N} \sum_{i=1}^N x_t^i B_s^i.$$

Compared to (3.2), χ_N correspond to an almost sure version of the integral of the response function; we shall prove that χ_N converges almost surely and that its limit is differentiable, hence showing the convergence in distribution of the response function.

Theorem 3.1. *Let $\psi(r) = \nu'(r) + r\nu''(r)$. Fixing any $T < \infty$, as $N \to \infty$ the random functions K_N and χ_N converge uniformly on $[0,T]^2$, almost surely and in L_p with respect to $\mathbf{x}_0$, $\mathbf{J}$ and $\mathbf{B}$ to non-random functions $\chi(s,t) = \int_0^t R(s,u)du$ and $K(s,t) = K(t,s)$. Further, $R(s,t) = 0$ for $t > s$, $R(s,s) = 1$, and for $s > t$ the absolutely continuous functions K, R and $K(s) = K(s,s)$ are the unique solution in the space of bounded, continuous functions, of the integro-differential equations*

$$\partial_s R(s,t) = -U'(K(s))R(s,t) + \beta^2 \int_t^s R(u,t)R(s,u)\nu''(K(s,u))du, \quad (3.5)$$

$$\partial_s K(s,t) = -U'(K(s))K(s,t) + \beta^2 \int_0^s K(u,t)R(s,u)\nu''(K(s,u))du$$

$$+ \beta^2 \int_0^t \nu'(K(s,u))R(t,u)du, \quad (3.6)$$

$$\partial_s K(s) = -2U'(K(s))K(s) + 1 + 2\beta^2 \int_0^s \psi(K(s,u))R(s,u)du, \quad (3.7)$$

with the initial condition $K(0) = \int x^2 d\mu_0(x)$.

Note that, setting $K(s) = 1$ and $\partial_s K(s) = 0$ in (3.7), while replacing $U'(K(s))$ in (3.5)–(3.7) by a time varying constant $z(s)$, corresponds to the hard spherical constraint of [30]. The limiting equations of [30] are thus recovered under the assumption that $K(s)$ converges as $s \to \infty$. Another way to recover exactly Cugliandolo–Kurchan's equations is to take a potential $U = L(x-1)^2 + x^m$ and consider the solution (K_L, R_L) of (3.5)–(3.7) with such a potential. Then we proved in [32] that these solutions converge as L goes to infinity towards the solution of the original Cugliandolo–Kurchan's equations which amount to take $K(s) = 1$ for all s and replace $U'(K(s))$ by

$$\frac{1}{2}\left(1 + 2\beta^2 \int_0^s \psi(K(s,u))R(s,u)du\right)$$

in the equations for $K(s,t)$ and $R(s,t)$.

Outline of the proof. There are mainly two difficulties to overcome in the p-spins model in comparison with the case where $p = 2$; the tightness issues (which cannot rely on any diagonalization of a matrix), and the derivation of the equations (the analysis of the equations for R_N being completely new since for instance they do not depend on the annealed law as studied in [17]):

– The first difficulty is that when $p > 2$, we cannot use any spectral analysis as before. However, for tightness issues, we can still rely on known properties of Gaussian processes [33], namely for any $p > 0$, if we set

$$X_N(\mathbf{x}) = \sum_{1 \le i_j \le N, 1 \le j \le p} J_{\{i_1, \cdots, i_p\}} x_{i_1}^1 x_{i_2}^2 x_{i_3}^3 \cdots x_{i_p}^p,$$

we know that

$$X_N^*(\rho) = \sup_{\|\mathbf{x}^i\| \le \rho} \frac{|X_N(\mathbf{x})|}{\sqrt{N}}$$

satisfies, for a constant $\kappa < \infty$, for all $t > 0$

$$\mathbb{P}[X_N^*(\rho) \ge \kappa + t] \le \exp(-Nt^2/\kappa). \tag{3.8}$$

This estimate replaces the control on the spectral radius of the matrix $\mathbf{J}$ in the proof of the case $p = 2$ so that we can obtain tightness of K_N and R_N^J almost surely and in expectation.

- The next step is to prove the self-averaging of the quantities (K_N, χ_N). To do that, we show that K_N and χ_N are Lipschitz functions of $\mathbf{J}, \mathbf{B}, \mathbf{x}_0$ on a set with overwhelming probability (namely where the Euclidean norms of $\mathbf{J}, \mathbf{B}, \mathbf{x}_0$ are well bounded). This again relies on Gaussian estimates due to (3.8) which allows to see the random field $\nabla H_{\mathbf{J}}^\gamma(\mathbf{x})$ as a smooth function of the empirical covariance of $\mathbf{x}$. Once this is done, we merely rely on standard concentration inequalities based on Herbst's argument (which we can do as far as the law of $(\mathbf{x}_0, \mathbf{J}, \mathbf{B})$ satisfies a coercive inequality such as log-Sobolev inequality).
- The last, and may be most painful, step is to derive the equations. This is much more difficult here because we have to deal with the response function. In particular, since we could not prove the tightness of the response function R_N directly, but rather of its integrated form χ_N, we have to do many integration by parts to express all the equations in terms of χ_N, resulting with a loss of simplicity of the proofs.

It can be seen by integration by parts that the set of equations satisfied by (K, R) is equivalent to the following set of equations where R has been replaced by χ (in what follows ν has been changed into $\beta^2\nu$ to lighten the notations);

$$K(s,t) = K(s,0) + \chi(s,t) + \int_0^t D(s,u)du, \tag{3.9}$$

$$\chi(s,t) = s \wedge t + \int_0^s E(u,t)du, \tag{3.10}$$

$$D(s,t) = -U'(K(t,t))K(t,s) - \int_0^{t \vee s} \nu'(K(t,u))D(s,u)du$$

$$- \int_0^{t \vee s} K(s,u)\nu''(K(t,u))D(t,u)du$$

$$+ K(s, t \vee s)\nu'(K(t \vee s, t)) - K(s,0)\nu'(K(0,t)), \tag{3.11}$$

$$E(s,t) = -U'(K(s,s))\chi(s,t) - \int_0^s \nu'(K(s,u))E(u,t)du$$

$$- \int_0^s \chi(u,t)\nu''(K(s,u))D(s,u)du$$

$$+ \chi(s,t)\nu'(K(s,s)) - \int_0^{t\wedge s} \nu'(K(s,u))du, \qquad (3.12)$$

Here, recalling we can prove that $t \to \chi(s,t)$ is differentiable and so $\chi(s,t) = \int_0^t R(s,u)du$. Let us show how we derive (3.9) as well as the equation satisfied by D.

Taking the scalar product by $\mathbf{x}_s$ in (3.3) (once integrated between times s and t), we obtain, with $G_s^i(\mathbf{x}) = \partial_{x_i} H_{\mathbf{J}}^\gamma(\mathbf{x})$,

$$K_N(s,t) = K_N(s,s) + \chi_N(s,t) - \int_s^t U'(K_N(u))K_N(s,u)du$$

$$- \beta \int_s^t \frac{1}{N} \sum_{i=1}^N G_u^i(\mathbf{x})x_s^i du.$$

Taking the expectation on both sides of this equality and using that the a priori self-averaging property $K_N(u) \approx \mathbb{E}[K_N(u)]$ we get, with $\bar{K}_N = \mathbb{E}[K_N]$ and $\bar{\chi}_N = \mathbb{E}[\chi_N]$

$$\bar{K}_N(s,t) \approx \bar{K}_N(s,s) + \bar{\chi}_N(s,t) - \int_s^t U'(\bar{K}_N(u))\bar{K}_N(s,u)du$$

$$+ \beta \int_s^t \mathbb{E}\left[\frac{1}{N}\sum_{i=1}^N G_u^i(\mathbf{x})x_s^i\right] du.$$

Thus, if we set

$$D_N(s,t) = -U'(\bar{K}_N(t))\bar{K}_N(s,t) + \beta\mathbb{E}\left[\frac{1}{N}\sum_{i=1}^N G_t^i(\mathbf{x})x_s^i\right],$$

$$\bar{K}_N(s,t) \approx \bar{K}_N(s,s) + \bar{\chi}_N(s,t) + \int_s^t D_N(s,u)du. \qquad (3.13)$$

Hence, we get (3.9) provided we can prove that D_N converges towards D solution of (3.11). The difficult part to estimate $D_N(s,t)$ is to see how

$$\mathbb{E}\left[\frac{1}{N}\sum_{i=1}^N G_u^i(\mathbf{x})x_\tau^i\right] \quad \text{for} \quad u \geq \tau \qquad (3.14)$$

can be expressed in terms of $\bar{K}_N$ and $\bar{\chi}_N$. To do so, we compute $V_{\tau,u}^i = \mathbb{E}[G_u^i(\mathbf{x})|\mathcal{F}_\tau]$ with $\mathcal{F}_\tau$ the canonical filtration. Let us denote Λ_τ^N the density of the law $\mathbb{P}^N_{\mathbf{x}_0,\mathbf{J}}$ of $(\mathbf{x}_t, t \leq \tau)$ with respect to the law of the solution with $\beta = 0$, $\mathbb{P}^N_{\mathbf{x}_0,0}$; By Girsanov formula, if G^i_s denotes in short $G^i(\mathbf{x}_s)$, we have the formula

$$\Lambda^N_\tau = \exp\left\{\sum_{i=1}^{N} \int_0^\tau G^i_s dW^i_s(\mathbf{x}) - \frac{1}{2}\sum_{i=1}^{N} \int_0^\tau (G^i_s)^2 ds\right\}. \qquad (3.15)$$

with $W^i_s(\mathbf{x}) = x^i_s - x^i_0 + \int_0^s U'(K_N(u))x^i_u du$. Hence, with $\tau \geq s$, for any bounded $\mathcal{F}_\tau$-measurable random variable Φ,

$$\mathbb{E}\left[G^i_s\Phi\right] = \mathbb{E}_{\mathbf{J}}\mathbb{E}_{\mathbb{P}^N_{\mathbf{x}_0,\mathbf{J}}}\left[G^i_s\Phi\right] = \mathbb{E}_{\mathbb{P}^N_{\mathbf{x}_0,0}}\left[\mathbb{E}_{\mathbf{J}}\left[G^i_s\Lambda^N_\tau\right]\Phi\right] = \mathbb{E}\left[\frac{\mathbb{E}_{\mathbf{J}}[G^i_s\Lambda^N_\tau]}{\mathbb{E}_{\mathbf{J}}[\Lambda^N_\tau]}\Phi\right], \qquad (3.16)$$

where the right-most identity is due to the change of measure formula $\mathbb{Q}^N_{\mathbf{x}_0} = \mathbb{E}_{\mathbf{J}}(\Lambda^N_\tau)\mathbb{P}^N_{\mathbf{x}_0,0}$ for the annealed law $\mathbb{Q}^N_{\mathbf{x}_0} = \mathbb{E}_{\mathbf{J}}\mathbb{P}^N_{\mathbf{x}_0,\mathbf{J}}$, restricted to $\mathcal{F}_\tau$. With (3.16) holding for all bounded $\mathcal{F}_\tau$-measurable Φ, it follows that:

$$V^i_s = \mathbb{E}\left[G^i_s|\mathcal{F}_\tau\right] = \frac{\mathbb{E}_{\mathbf{J}}[G^i_s\Lambda^N_\tau]}{\mathbb{E}_{\mathbf{J}}[\Lambda^N_\tau]}.$$

Now, by standard Gaussian computations (here $\mathbf{x}$ is fixed and we only integrate over the Gaussian variables), we see that if $k^{ij}_{uv} = \mathbb{E}_{\mathbf{J}}[G^i(\mathbf{x}_u)G^j(\mathbf{x}_v)]$ and k_τ is the integral operator so that for $f \in L^2([0,\tau])^N$, $u \leq \tau$

$$[k_\tau f]^i_u = \sum_{j=1}^{N} \int_0^\tau k^{ij}_{uv} f^j_v dv,$$

$$V^i_s + [k_\tau V]^i_s = [k_\tau \circ dW]^i_s. \qquad (3.17)$$

Here, dW is seen as a semi-martingale and $k_\tau \circ dW$ as a stochastic integral. One can easily compute $k^{ij}_{tu}(\mathbf{x})$;

$$k^{ij}_{ts}(\mathbf{x}) := \mathbb{E}_{\mathbf{J}}\left[G^i(\mathbf{x}_t)G^j(\mathbf{x}_s)\right] = \frac{x^j_t x^i_s}{N}\nu''(K_N(s,t)) + \mathbf{1}_{i=j}\nu'(K_N(s,t)). \qquad (3.18)$$

Writing for $\tau = \max\{u, s\}$

$$\mathbb{E}\left[\frac{1}{N}\sum_{i=1}^{N} G^i_u(\mathbf{x})x^i_s\right] = \mathbb{E}\left[\frac{1}{N}\sum_{i=1}^{N} V^i_s x^i_s\right],$$

the rest of the proof becomes (rather painful) algebra based on (3.17) and (3.18).

The equation for R is slightly more involved, in particular to obtain a formula for

$$\mathbb{E}\left[\frac{1}{N}\sum_{i=1}^{N} G^i_u B^i_s\right] \quad \text{for } s \leq u.$$

I refer the reader to [23] for details.

Once (3.5), (3.6) have been derived, the question at stake is to understand their limiting behaviour as the time parameters t, s go to infinity.

Note that for $p = 2$, i.e. $\nu(r) = r^2/2$, we get from (3.5) the autonomous equation

$$\partial_s H(s,t) = \beta^2 \int_t^s H(u,t)H(s,u)du, \quad H(t,t) = 1,$$

for $H(s,t) = R(s,t)\exp(\int_t^s U'(K(u))du)$, whose unique solution is the Laplace transform of the semi-circle probability measure, $H(s,t) = (2\pi)^{-1}\int_{-2}^2 \exp(\beta(s-t)x)\sqrt{4-x^2}dx$, evaluated at $\beta(s-t)$. Plugging this expression in (3.5) and (3.6), we recover the limiting equation of the previous section after some integrations by parts. A natural question was to wonder whether the fact that the response function was related to the Laplace transform of the semi-circular law in the case $\nu(r) = r^2/2$ could be in some way generalized to other ν. This question was addressed in [34] where the following surprising remark was made; let $k(t,s) = \beta^2\nu''(C(s,t))$ and consider H as earlier. Then, H satisfies the non-linear Kraichnan equation

$$\partial_s H(s,t) = \int_t^s H(s,u)H(u,t)k(s,u)du, \quad s \geq t \qquad (3.19)$$

with the boundary condition $H(t,t) = 1$. Such an equation already appeared in the work of Kraichnan [35] as a first term in a perturbative method to analyze quantum-mechanical, turbulence or disordered problems. It then appeared in relation with differential equations for non-commutative processes in the works of Frisch and Bourret [36] and Neu and Speicher [37]. This last interpretation goes as follows; Let $(L_t)_{t \geq 0}$ be a process in a von Neumann algebra $\mathcal{A}$ equipped with a tracial state ϕ. I assume that L is a centered semi-circular process with covariance kernel k, usually constructed on the full Fock space (see e.g. [38]). In a more intuitive way, L can be constructed as the limit of self-adjoint large random matrices $(L_t^N)_{t \geq 0}$ with entries $\{(L_t^N)_{ij}, 1 \leq i \leq j \leq N\}$ which are independent Gaussian processes with covariance $N^{-1}k$. This limit has to be understood in the weak sense that for any integer number n, any times $(t_1, t_2, \cdots t_n) \in (\mathbb{R}^+)^n$,

$$\lim_{N \to \infty} \frac{1}{N}\mathrm{tr}\left(L_{t_1}^N L_{t_2}^N \cdots L_{t_n}^N\right) = \phi\left(L_{t_1} L_{t_2} \cdots L_{t_n}\right),$$

where tr denotes the unnormalized trace of matrices. Consider then the family of operators $\mathbf{X}_{s,t}$ satisfying the linear differential equation

$$\partial_s \mathbf{X}_{s,t} = L_s \mathbf{X}_{s,t}, \quad s > t,$$

with boundary data $\mathbf{X}_{t,t} = 1$, on the full Fock space. Then, it was shown in [36] that $H(s,t) = \phi(\mathbf{X}_{s,t})$ satisfies Kraichnan's equation (3.19), equation which can easily be seen to have a unique solution (see [34]). In terms of matrices,

$H(s, t)$ can be seen to be the limit as N goes to infinity of trace of the solution $\mathbf{X}_{s,t}^N$ of the random linear differential equation $\partial_s \mathbf{X}_{s,t}^N = L_s^N \mathbf{X}_{s,t}^N$, $s > t$ such that $\mathbf{X}_{t,t}^N = I$.

In the case where $\nu(r) = r^2/2$, we recover the fact that H is the Laplace transform of the semi-circular law since then $L_s^N = \beta \mathbf{J}$ does not depend on s so that $\mathbf{X}_{s,t}^N = e^{(s-t)\beta \mathbf{J}}$.

In the general case, the behaviour of H of course depends on that of $k = \beta^2 \nu''(C)$ in a rather complicated way. Using the intuition coming from free probability, we can derive a formula for H in terms of k by using the analogue of Wick formula in free probability;

$$H(s,t) = \sum_{n \geq 0} \int_{t \leq t_1 \cdots \leq t_{2n} \leq s} \sum_{\sigma \in NC_n} \prod_{i \in \mathrm{cro}(\sigma)} k(t_i, t_{\sigma(i)}) dt_1 \cdots dt_{2n} \qquad (3.20)$$

where NC_n denotes the set of involutions of $\{1, \cdots, 2n\}$ without fixed points and without crossings and where $\mathrm{cro}(\sigma)$ is defined to be the set of indices $1 \leq i \leq 2n$ such that $i < \sigma(i)$. $\sigma \in NC_n$ when the situation $i < j < \sigma(i) < \sigma(j)$ does not occur.

In [34], Mazza and I studied the long-time behaviour of H being given that of k to try to get some intuition about the equation governing the response function. Roughly speaking, we found out that if k is stationary, under rather general assumptions

1. If $\lim_{u \to \infty} k(u) = 0$, there exists $\lambda_c(H) > 0$ such that

$$\exp(-\lambda_c(H)t)H(t) \sim \frac{1}{2A}, \quad t \to +\infty,$$

2. If $\lim_{u \to \infty} k(u) > 0$, there exists $\lambda_c(H) > 0$ such that

$$e^{-\lambda_c(H)t} H(t) \sim At^{-3/2}, \quad t \to +\infty,$$

for some positive constant A.

We cannot in general compute the Lyapounov exponent $\lambda_c(H)$ except in the case where $k(u) = ce^{-\delta u}$. In this case, $\lambda_c(H)$ appears to be the smallest zero of a Bessel function. The stationary case is much simpler than the general case because then one easily derives a formula for the Fourier transform of H, on which analysis can be developped. However, the first-order asymptotics (the $\lambda_c(H)$) can often be studied (see [34]).

Hence, we see from this study that the system (3.5), (3.6) is a rather complicated one. However, in a work in progress with Dembo and Mazza [32] we could study the asymptotic behaviour of (R, C) for sufficiently small β. To do that, we first proved that the assumptions made in [34] about the fact that $k = \beta^2 \nu''(C)$ is uniformly bounded and non-negative is true;

Lemma 3.2. *At all temperatures, the unique solution (R, K) to (27)–(29) are non-negative functions. K is uniformly bounded above and below by positive constants.*

This fact then can be used to show that

Theorem 3.3. *Assume that β is small enough and $\nu''(0) = 0$. Then, there exists $\delta_\beta > 0$ and $A > 0$ so that for all times t, s*

$$R(s,t) \leq Ae^{-\delta_\beta|s-t|}, \quad K(s,t) \leq Ae^{-\delta_\beta|s-t|}.$$

In this very high temperature regime, we can also show that the dynamics become stationary in the sense that

$$\lim_{t\to\infty} R(t,t+s) = R_{FDT}(s) \quad \lim_{t\to\infty} K(t,t+s) = K_{FDT}(s) \quad \lim_{t\to\infty} K(t) = K(\infty).$$

Further, we can describe the integro-differential equations satisfied by $(R_{FDT}(s), K_{FDT}(s))_{s\geq 0}$ which are the so-called FDT-equations introduced first by Cugliandolo and Kurchan [30]; $R_{FDT}(s) = -2\partial_s K_{FDT}(s)$ and K_{FDT} is the unique solution of the equation

$$X'(s) = -\gamma X(s) - 2\beta^2 \int_0^s \nu'(X(v))X'(s-v)dv. \tag{3.21}$$

with $X(1) = 1$ and $\gamma = f'(X(0)) - 2\beta^2\nu'(X(0)) = -1/2$.

4 Future problems

When the question of aging is adressed, it raises more questions and open problems than solutions. Let us describe a short list later;

- The most immediate problem on which I am still working with Ben Arous, Dembo, and Mazza is the understanding of the FDT equations at all temperature and more generally the study of the limiting p-spins dynamics in the low-temperature region. In particular, it would be very nice to see whether the properties $(\nu'')^{-\frac{1}{2}}$ governs, as for the statics, the type of dynamical phase transition of the model. This leads actually to the much deeper following problem.
- What are the relations between the dynamical phase transitions and the static phase transition ? This relation seems very obsceure since it is known (according to a private communication of Cugliandolo) that glasses dynamics are likely to stay stuck very far from their ground states, and so should not "see" the Gibbs measure. However, in many models, the dynamical phase transition seems to mimic properties of the statics (see the original Sherrington–Kirkpatrick model) and below the dynamical phase transition, the FDT solutions seems to be related with the solutions starting from the invariant measure. In this direction, spherical models seems to be a good set of models since their static has been recently analyzed by Talagrand [20] and the limiting dynamics introduced Cugliandolo and Kurchan have been validated by [23].

- It would be interesting to consider true dynamics on the sphere, as far as one would like to understand the dynamics for the hard-sphere equilibrium measure studied by Talagrand [20]. It is actually not clear at all that it would give similar dynamical equations.
- There are many other systems which should age, such as granular materials. It would be interesting to understand how aging appear in other materials. In physics, a domain growth approach have been proposed to explain aging phenomenon in some models (see [5]). It would be worth to understand it.
- Another set of questions appears when the interaction is short range or given by a Kac model; for the equilibrium Gibbs measure, it is still under discussion even in physics, what should be the picture and whether the ultrametric picture holds. The limiting dynamics for the short-range model of the Sherrington–Kirkpatrick were derived in [39], and it is likely that the generalization to spherical models could be derived by combining the techniques presented in this survey and those of [39].

Acknowledgments

I am very grateful to A. Dembo for his careful reading and useful comments on a preliminary version of this review.

References

1. Ruelle D.; *Statistical mechanics: rigorous results*, Benjamin, Amsterdam (1969)
2. Nattermann T., Scheidi S.; Vortex glass phases in type-II superconductors, *Adv. Phys.* **49**, 607 (2000)
3. Biljakovic K., Lasjaunias J.C., Monceau P.; Aging effects and non exponential energy relaxations in charge-density wave systems, *Phys. Rev. Lett.* **62** 1512 (1989)
4. Bouchaud J.P., Granular media: some ideas from statistical physics; arxiv.org/cond-mat/0211196
5. Cugliandolo L.; Dynamics of glassy systems; http://xxx.lanl.gov/abs/cond-mat/0210312
6. Ben Arous G., Cerny J.; Bouchaud's model exhibits two aging regime in dimension one, *Ann. Appl. Probab.* **15** 1161–1192 (2005)
7. Ben Arous G., Cerny J., Mountford T.; Aging for Bouchaud's model in dimension 2, *Probab. Theo. Relat. Fields* **134** 1–43 (2006)
8. Dembo A., Guionnet A., Zeitouni O.; Aging properties of Sinai's random walk in random environment, *XXX preprint archive*, math.PR/0105215 (2001) published in ZEITOUNI O.; Random walks in random environment, *Lecture Notes in Mathematics* **1837** 189–312 (2004)
9. Cheliotis D.; Diffusion in random environment and the renewal Theorem *Ann. Probab.* **33** 1760–1781 (2005)
10. Dembo A., Deuschel J.D.; Aging for interacting diffusion processes, *preprint* (2006)

11. Ben Arous G., Bovier A., Gayrard V.; Glauber dynamics of the random energy model I. Metastable motion on the extreme states. *Commun. Math. Phys.* **235** 379–425 (2003)

12. Ben Arous G, Bovier A., Gayrard V.; Glauber dynamics of the random energy model II; Aging below the critical temperature. *Commun. Math. Phys.* **236** 1–54 (2003)

13. Grunwald M.; Sanov results for Glauber spin-glass dynamics. *Probab. Theo. Relat. Fields* **106**, 187–232 (1996)

14. Sompolinsky H., Zippelius A.; Phys. Rev. Lett. **47** 359 (1981)

15. Mezard M., Parisi G., Virasoro M.; Spin glass theory and beyond. *World Scientific Lecture Notes in Physic* (1987)

16. Ben Arous G., Guionnet A.; Symmetric Langevin spin glass dynamics. *Ann. Probab.* **25** 1367–1422 (1997)

17. Ben Arous G., Guionnet A.; Sherrington Kirkpatrick spin glass dynamics *Progr. Prob.* **41** 323–353 (1998)

18. Guionnet A.; Annealed and quenched propagation of chaos for Langevin spin glass dynamics. *Probab. Theo. Relat. Fields* **109**, 183–215 (1997)

19. Guionnet A.; Non Markovian limit diffusions and spin glasses *Fields Inst. Commun.* **34** 65–74 (2002)

20. Talagrand M.; Free energy of the spherical mean field model, *Preprint* (2004)

21. Cugliandolo L., Dean D.S.; Full dynamical solution for a spherical spin-glass model. *J. Phys.* A **28** 4213 (1995)

22. Ben Arous G., Dembo A., Guionnet A.; Aging of spherical spin glasses. *Probab. Theo. Relat. Fields* **120** 1–67 (2001)

23. Ben Arous G., Dembo A., Guionnet A.; Cugliandolo-Kurchan equations for dynamics of Spin-Glasses, to appear in *Probab. Theo. Relat. Fields* (2006)

24. Ledoux M.; The concentration of measure phenomenon, *Am. Math. Soc.* providence (2001)

25. Ané C., Blachère S., Khafi D., Fougerères P., Gentil I., Malrieu F., Roberto C., Scheffer G.; *Sur les inégalités de Sobolev logarithmique* Panoramas et synthèse, **11**, 120, 121 (2000)

26. Bai, Z.D., Yin Y.Q.; Necessary and sufficient conditions for almost sure convergence of the largest eigenvalue of a Wigner matrix. *Ann. Probab.* **16** 1729–1741 (1988)

27. Füredi Z., Komlós J.; The eigenvalues of random symmetric matrices, *Combinatorica* **1**, 233–241 (1981)

28. Ledoux M.; Isoperimetry and Gaussian analysis, Lectures on probability theory and statistics. Lectures from the 24th Saint-Flour Summer School held July 7–23, 1994. Edited by P. Bernard. *Lecture Notes in Mathematics, 1648.* Springer Heidelberg Newyork, Berlin, (165–294), 1996

29. Wigner E.; On the distribution of the roots of certain symmetric matrices. *Ann. Math.* **67**, 325–327 (1958)

30. Cugliandolo L., Kurchan J.; Analytical solution of the off-equilibrium dynamics of a long range spin-glass model. *Phys. Rev. Lett.* **71** 173 (1993)

31. Cugliandolo L., Kurchan J.; On the out of Equilibrium relaxation of the Sherrington-Kirkpatrick model. *J. Phys.* A **27** 5749 (1994)

32. Dembo A., Guionnet A., Mazza C.; Dynamics for spherical p-spins at high temperature, Preprint (2006)

33. Ledoux M., Talagrand M.; Probability in banach spaces. *Ergebnisse der Mathematik 23, Springer, Berline Heidelberg Newyork* (1991)

34. Guionnet A., Mazza C.; Long time behaviour of non-commutative processes solution of a linear differential equation. *Prob. Theo. Relat. Fields* **131** 493–518 (2005)
35. Kraichnan R.; Dynamics of nonlinear stochastic systems. *J. Math. Phys.* **2**, 124 (1961)
36. Frisch U., Bourret R.; Parastochastics *J. Math. Phys.* **11**, 364 (1970)
37. Neu P., Speicher R.; A self-consistent master equation and a new kind of cumulants. *Z. Phys. B* **92**, 399 (1993)
38. Voiculescu D.; Lectures on free probability. in *Lectures Notes in Mathematics 1738*, Springer, Berlin Heidelberg Newyork (2000)
39. Ben Arous G., Sortais M.; Large deviations in the Langevin dynamics of a short range spin-glass. *Bernouilli* **9** 921–954 (2003)

Local vs. Global Variables
for Spin Glasses

Charles M. Newman and Daniel L. Stein

Courant Institute of Mathematical Sciences, New York University
251 Mercer Street, New York, NY 10012, USA
e-mail: newman@cims.nyu.edu
and

Departments of Physics and Mathematics, University of Arizona
1118 E. 4th Street, PO Box 210081, Tucson, AZ 85721, USA
e-mail: dls@physics.arizona.edu

Summary. We discuss a framework for understanding why spin glasses differ so remarkably from homogeneous systems like ferromagnets, in the context of the sharply divergent low temperature behavior of short- and infinite-range versions of the same model. Our analysis is grounded in understanding the distinction between two broad classes of thermodynamic variables–those that describe the *global* features of a macroscopic system, and those that describe, or are sensitive to, its *local* features. In homogeneous systems both variables generally behave similarly, but this is not at all so in spin glasses. In much of the literature these two different classes of variables were commingled and confused. By analyzing their quite different behaviors in finite- and infinite-range spin glass models, we see the fundamental reason why the two systems possess very different types of low-temperature phases. In so doing, we also reconcile apparent discrepancies between the infinite-volume limit and the behavior of large, finite volumes, and provide tools for understanding inhomogeneous systems in a wide array of contexts. We further propose a set of "global variables" that are definable and sensible for both short-range and infinite-range spin glasses, and allow a meaningful basis for comparison of their low-temperature properties.

Key words: Spin glass, Edwards–Anderson model, Sherrington–Kirkpatrick model, Replica symmetry breaking, Mean–field theory, Pure states, Metastates, Domain walls, Interfaces

1 Local Variables and Thermodynamic Limits

In recent years, attempts have been made [1, 2] to draw a distinction between the thermodynamic limit as a "mathematical tool" of limited physical relevance, and the physical behavior of large, finite systems, the real-world objects of study. In this note we discuss why this distinction is spurious, and show through several examples that this "tool" is useful precisely *for* determining the behavior of large, finite systems. We will also discuss why this attempted distinction has caused confusion in the case of spin glasses, and how resolving it introduces some important new physics. For ease of discussion, we confine ourselves throughout to Ising spin systems.

One of the incongruities of the "thermodynamic limit vs. finite volume" debate is that true thermodynamic states are in fact measures of the *local* properties of a macroscopic system, while discussions of finite-volume properties – at least in the context of infinite-range spin glasses – focus entirely on *global* quantities.

Of course, traditional thermodynamics (as opposed to statistical mechanics) is entirely a study of global properties of macroscopic systems. Quantities like energy or magnetization are collective properties of *all* of the individual spins in a given finite- or infinite-volume configuration. Other global measures cannot be discussed in terms of individual spin configurations but rather are meaningful only in the context of a Gibbs distribution (also known as Gibbs state or thermodynamic state – we will use these terms interchangeably throughout). Entropy is the obvious example of such a variable. All of these together – energy, entropy, magnetization, and the various free energies associated with them – convey only coarse information about a system (though still extremely valuable).

What they convey little information about is the actual spatial structure of a state, or of the relationships among different states. Even a quantity like the staggered magnetization, which gives some information about spatial structure, does not shed significant light on local properties.

But for real systems one often does want information about local properties. In order to analyze local spatial and temporal structures one generally needs to employ *local thermodynamic variables* – for example, $1-, 2-, \ldots, n$-point correlation functions. In fact, these functions taken altogether convey *all* of the information that can be known about that state in equilibrium – a Gibbs state is a specification of all possible n-point (spatial) correlation functions, for every positive integer n.

Alternatively, one can define a thermodynamic state either as a convergent sequence (or subsequence) of finite-volume states as volume tends to infinity, or else intrinsically through the DLR equations [3]. But we will avoid a technical discussion here in the interest of keeping the discussion focused on physical objects. We henceforth assume familiarity with concepts such as thermodynamic mixed state and thermodynamic pure state, which have been used extensively throughout much of the spin glass literature, and refer the reader who wishes to learn more to Sect. 4 of [4].

Are there any *global* quantities that say something about the spatial structure of a state? The answer is yes, and one in particular has proven to be very useful in comparing different pure state structures in the infinite-range

Sherrington–Kirkpatrick (SK) model [5] of a spin glass. The SK Hamiltonian (in volume N) is

$$\mathcal{H}_N = -\left(1\big/\sqrt{N}\right) \sum_{1 \leq i < j \leq N} J_{ij}\sigma_i\sigma_j \tag{1.1}$$

where the couplings J_{ij} are independent, identically distributed random variables chosen, e.g., from a Gaussian distribution with zero mean and variance one; the $1/\sqrt{N}$ scaling ensures a sensible thermodynamic limit for free energy per spin and other thermodynamic quantities.

In a series of papers, Parisi and collaborators [6–9] proposed, and worked out the consequences of, an extraordinary *ansatz* for the nature of the low-temperature phase of the SK model. Following the mathematical procedures underlying the solution, it came to be known as *replica symmetry breaking* (RSB). The starting point of the Parisi solution was that the low-temperature spin glass phase comprised not just a single spin-reversed pair of states, but rather "infinitely many pure thermodynamic states" [7], not related by any simple symmetry transformations.

But how were they related? To answer this question, Parisi introduced a *global* quantity–exactly of the sort we were just asking about – to quantify such relationships. The actual notion of pure "state" in the SK model is problematic, as discussed, e.g., in [4, 10–12]. We will ignore that problem for now, though, and assume that somehow two SK pure states α and β have been defined. Then their overlap $q_{\alpha\beta}$ is defined as

$$q_{\alpha\beta} = \frac{1}{N} \sum_{i=1}^{N} \langle\sigma_i\rangle_\alpha \langle\sigma_i\rangle_\beta \,, \tag{1.2}$$

where $\langle\cdot\rangle_\alpha$ is a thermal average in pure state α, and dependence on $\mathcal{J}$ and T has been suppressed. So $q_{\alpha\beta}$ is a quantity measuring the similarity between states α and β.

We noted above that quantities referring to individual pure states are problematic in the SK model, since there is no known procedure for constructing such states in a well-defined way. However, what is really of interest is the *distribution* of overlaps, which *can* be sensibly defined by using the finite-N Gibbs state. The overlap distribution is constructed by choosing, at fixed N and T, two of the many pure states present in the Gibbs state. The probability that their overlap lies between q and $q + dq$ is then given by the quantity $P_{\mathcal{J}}(q)dq$, where

$$P_{\mathcal{J}}(q) = \sum_\alpha \sum_\beta W_{\mathcal{J}}^\alpha W_{\mathcal{J}}^\beta \delta(q - q_{\alpha\beta}) \,. \tag{1.3}$$

As before, we suppress the dependence on T and N for ease of notation. The average $P(q)$ of $P_{\mathcal{J}}(q)$ over the disorder distribution is commonly referred to as the *Parisi overlap distribution*, and serves as an order parameter for the SK model.

Because there is no spatial structure in the infinite-range model, the overlap function does seem to capture the essential relations among the different states. However, it might already be noticed that such a global quantity would miss important information in short-range models – assuming that such models also have many pure states. There is no information in $P_{\mathcal{J}}(q)$ about local correlations. This is acceptable, even desirable, in an infinite-range model such as SK which has no geometric structure, measure of distance, or notion of locality or neighbor. But all of these are well-defined objects in short-range models, and carry a great deal of information about any state, pair of states, or collection of many states. This is one of the sources of the difficulties one encounters (see, e.g., [11]) when attempting to apply conclusions to short-range spin glasses that were derived for the SK model.

2 Nearest-Neighbor Ising Ferromagnets

To illustrate some of these ideas in a simple context, consider the uniform nearest-neighbor Ising ferromagnet on $\mathbf{Z}^d$, with Hamiltonian

$$\mathcal{H} = - \sum_{\substack{x,y \\ |x-y|=1}} \sigma_x \sigma_y . \tag{2.1}$$

It is natural in models such as this to take periodic or free boundary conditions when considering the finite-volume Gibbs state. In any fixed $d \geq 2$, consider for $T < T_c$ a sequence of volumes Λ_L (for specificity, L^d cubes centered at the origin) tending to infinity with, for example, periodic b.c.'s. The Gibbs state – which, as emphasized in Sect. 1, describes behavior of the local spin variables – converges in the infinite-volume limit to the symmetric mixture of a pure plus state with $\langle \sigma_x \rangle > 0$ (which, because of translation symmetry, is independent of x) and a pure minus state with $\langle \sigma_x \rangle < 0$. In a similar way, one could choose to study instead a global variable, say the magnetization per spin $M_L = |\Lambda_L|^{-1} \sum_{x \in \Lambda_L} \sigma_x$. It is easy to show that this variable has a distribution that converges in the limit to a symmetric mixture of δ-functions at $\pm \langle \sigma_x \rangle$. In this case descriptions of the system in terms of both local and global variables are interesting, and more to the point, they agree.

But even in the simple case of the uniform, nearest-neighbor Ising ferromagnet this need not always be true. Consider "Dobrushin boundary conditions" [13]. These are b.c.'s in which the boundary spins above the "equator" (a plane or hyperplane parallel to two opposing faces of Λ_L and cutting it essentially in half by passing just above the origin) are chosen to be plus and the boundary spins at and below the equator are minus. Boundary conditions such as these are useful for studying interface structure in spin models.

Now (below the roughening temperature) translation-invariance is lost (in one direction), and the local and global variables disagree. The Gibbs state is one where $\langle \sigma_z \rangle > 0$ for $z > 0$ (taking $z = 0$ to be the equator) and $\langle \sigma_z \rangle < 0$ when $z \leq 0$. The magnitude of $\langle \sigma_z \rangle$ will depend on the value of z (though it remains independent of the coordinates in all transverse directions). Moreover for $0 < T < T_c$ there is additional dimension-dependence of the behavior of the local variables (related to the roughening transition) which we will not discuss here.

The magnetization global variable is no longer even interesting. Its distribution converges to a δ-function at 0 at all temperatures in all dimensions. It therefore conveys very little information about the nature of the state. One could instead choose a more appropriate global variable that better matches the boundary conditions, e.g., an order parameter such as

$$\widetilde{M} = \lim_{L \to \infty} |\Lambda_L|^{-1} \sum_{x \in \Lambda_L} g(x) \langle \sigma_x \rangle \tag{2.2}$$

where $g(x) = +1$ if x is above the equator and -1 if below.

One might also consider a sort of "quasiglobal" variable, that looks at block magnetizations in blocks that are large compared to the unit lattice spacing but small compared to entire system size L. One could then examine the "spatial" distribution of the block magnetization as the location of the block varies through the system. Above T_c this is simply a δ-function at zero, but below T_c one gets a symmetric mixture of δ-functions at $\pm \lim_{z \to \infty} \langle \sigma_z \rangle$ for all d. This is still not as sensitive as the actual Gibbs state, which can distinguish between the rough and nonrough interfaces (e.g., below T_c in $d = 2$ compared to $T < T_R$ in $d = 3$, where T_R is the 3D roughening temperature) by having a different expression for the limiting Gibbs state in the two cases.

3 Finite-Range and Infinite-Range Spin Glasses

In the case of spin glass models, we have found [10, 11, 14–18] a much sharper disparity between finite- and infinite-range models than is the case for any homogeneous statistical mechanical model of which we are aware. Consider first the Edwards–Anderson (EA) nearest-nieghbor model [19] on $\mathbf{Z}^d$. Its Hamiltonian in zero external field is given by

$$\mathcal{H} = - \sum_{\substack{x,y \\ |x-y|=1}} J_{xy} \sigma_x \sigma_y \,, \tag{3.1}$$

where the nearest-neighbor couplings J_{xy} are defined in exactly the same way as the J_{ij} in the SK Hamiltonian (3.1). In this model thermodynamic pure, mixed, and ground states are standard, well-defined (see, e.g., [4,18]) objects, constructed according to well-established prescriptions of statistical mechanics [3,20–25]. Local thermodynamic variables therefore convey in principle all of the essential information about any state.

Global variables such as Parisi overlap functions can be defined for the EA model as well, but are now very prescription-dependent: for the same system, very different overlap functions can be obtained through use of different boundary conditions, or by changing the order of taking the thermodynamic limit and breaking the replica symmetry. Because of this, they may not convey reliable information about the number of states or the relationships among them. For a detailed review of these issues and problems, see [4].

Turning to the SK model, we find that a unique situation arises. Pure states are in principle defined for a fixed realization of *all* of the couplings; but in the SK model the *physical* couplings $J_{ij}/\sqrt{N}$ scale to zero as $N \to \infty$. As a result, there does not now exist any known way of constructing thermodynamic pure states in an SK spin glass. It has been proposed [26, 27] that one way of defining such objects is through the use of a modified "clustering" property: if α denotes a putative pure state in the SK model, then one can demand it satisfy:

$$\langle\sigma_i\sigma_j\rangle_\alpha - \langle\sigma_i\rangle_\alpha\langle\sigma_j\rangle_\alpha \to 0 \qquad \text{as} \quad N \to \infty, \tag{3.2}$$

for any fixed pair i, j, in analogy with the clustering property obeyed by "ordinary" pure states in conventional statistical mechanics. At this time though, there exists no known operational way to construct such an α as appears in (3.2). But even if such a construction were available, the definition of pure states through (3.2) leads to bizarre conclusions in the SK model (see [11,28]).

Are these problems simply a consequence of infinite-range interactions? No, because they are absent in the Curie–Weiss model of the uniform ferromagnet. Though physical couplings scale to zero there also, they "reinforce" each other, being nonrandom, so one may still talk about positive and negative magnetization states – in analogy with what one sees in finite-range models – in the $N \to \infty$ limit (of course, one can no longer talk of interface states). So the unique behavior of the SK model arises (at the least) from the combination of *two* properties: coupling magnitudes scaling to zero as $N \to \infty$, and quenched disorder in their signs. (Some success has been achieved in defining states in mean-field Hopfield models – see, e.g., [29–31] and references therein – where the correct order parameters are known *a priori*.)

Although individual pure states have not so far been (and perhaps cannot even in principle be) constructed for the SK model, we have nevertheless proposed methods, based on chaotic size dependence [32], that can detect the *presence* of multiple pure states. One can then examine the nature of objects that are analogous to states and that are defined through local variables, using the usual prescriptions of statistical mechanics. However, an analysis [11] of the properties of these state-like objects shows that they behave in completely unsatisfactory – in fact, absurd – ways. For example, using the traditional definition of a ground state – or equivalently, the modified clustering property of (3.2) – one can prove that (for almost every fixed coupling realization $\mathcal{J}$) *every* infinite-volume spin configuration is a ground state. That is, as N increases, any fixed finite set of correlation functions cycles through all of its possible sign configurations infinitely many times. Of course, this cannot happen in short-range spin glasses in any dimension, nor in any other statistical mechanical model based on any sort of physical system.

The upshot is that global variables (like overlap distributions for the whole system) capture interesting phenomena in the SK model, while local variables are not so interesting there; in fact, their use can even be dangerous in drawing conclusions about realistic spin glass models. This may be because the SK model itself is a priori a global (or at least nonlocal) model, and does not lend itself even in principle to any sort of local analysis.

A large body of evidence compiled by the authors [10, 11, 14–18] shows that local variables and states in the EA model do not behave anything like those in the SK model. The same conclusion applies to global variables in the EA model constructed in close analogy with those from the SK model – that is, in a way intended to convey information about states. This will be further discussed in [28]. Consequently, we expect that attempts to derive conclusions about the EA model in terms of local properties (i.e., pure state behavior) from the global behavior of the SK model will not work.

4 A New Global Order Parameter for Spin Glasses?

In this section we consider an interesting speculative question motivated by a comment of Bovier [33] that the usual description of Gibbs measures for short-range models is inadequate for mean field spin glasses. We have presented a large body of evidence that any quantity, describing spin glass properties and that is derived from or based on properties of pure states, cannot connect the behavior of short-range and infinite-range spin glasses. But can one construct a new type of global variable that is meaningful for both short-range and infinite-range spin glass models, and allows a direct comparison of their properties? One obvious candidate is a "global" overlap function not related to pure states; that is, rather than computing $P_L(q)$ in a "window" [4, 17, 18] far from $\partial \Lambda_L$, one would compute it in the entire volume Λ_L. As discussed in [4, 17, 18], the resulting quantity may be unrelated to pure state structure. An analogous situation is the ferromagnet *above* the roughening temperature (but below T_c). Even though there are no domain wall states, employing Dobrushin boundary conditions (see Sect. 2) will generate spin configurations that on very large scales (say, of order L) look almost indistinguishable from those belonging to domain wall states. But it is unclear whether doing this generates any useful or nonobvious information.

Similarly (see [17] for a more detailed discussion) there is reason to doubt whether a global overlap distribution would be any more useful. In the SK model boundary conditions are not an issue; but in short-range models, overlaps are potentially very sensitive to them. One consequence [4, 34] of this sensitivity is that spin overlap functions are unreliable indicators of pure state multiplicity: they can possess a trivial structure (cf. Fig. 1d in the companion paper in this volume [28]) in systems with infinitely many pure states, and a more complicated structure [34] in systems with only a single pure state. In particular, spin overlap functions computed in systems with quenched disorder can easily display complicated and nonphysical behavior that simply reflects the "mismatch" between the boundary condition choice and the local coupling variables. Consequently, nontrivial spin overlaps invariably generated by a change in boundary conditions (e.g., by switching from periodic to antiperiodic in spin glasses) may carry no more significance than that, say, in the 2D Ising ferromagnet (for $0 < T < T_c$) generated by Dobrushin boundary conditions. These considerations make it difficult to believe that a spin overlap variable – even when confined to a window – is likely to uncover generally useful information in short-range spin glasses.

Nevertheless, one can conceive and construct different, and possibly more useful, global quantities that can provide useful statistical mechanical information. Of greater interest, they *can* be used in principle to compare and contrast physically meaningful behaviors of short-range and infinite-range spin glasses. We propose here that one such set of quantities are those related to *interfaces* between spin configurations drawn at random from finite-volume Gibbs distributions at fixed L and T.

The interface between two such configurations σ^L and σ'^L is simply the set of all couplings that are satisfied in one of the two configurations but not the other. An interface separates regions where the spins in the two configurations agree from those where they disagree. We propose that the global variables of interest are those – in particular, density and energy – characterizing the physical properties of these interfaces.

The use of interfaces as a probe of spin glass structure is not new. Our purpose here is to argue, based on the overall approach described in this note, that their properties provide a significantly more natural and useful set of global spin glass variables than the spin overlap function.

The interface density in particular has of course been studied in earlier papers [2, 18, 35–39], as the edge overlap $q_e^{(L)}(\sigma, \sigma')$ between σ^L and σ'^L:

$$q_e^{(L)}(\sigma, \sigma') = N_b^{-1} \sum_{\substack{x,y \in \Lambda_L \\ |x-y|=1}} \sigma_x \sigma_y \sigma'_x \sigma'_y \tag{4.1}$$

where N_b denotes the number of bonds inside Λ_L. In the SK model, the edge overlap is defined similarly, except that the sum runs over every pair of spins.

In the SK model, there is a trivial relationship between the edge and spin overlaps. Consider two spin configurations σ and σ' in an N-spin system; their spin overlap is defined in the usual way as

$$q_s^{(N)}(\sigma, \sigma') = N^{-1} \sum_{i=1}^{N} \sigma_i \sigma_i'. \tag{4.2}$$

It is easy to see that

$$q_e^{(SK)}(\sigma, \sigma') = \left(q_s^{(SK)}(\sigma, \sigma')\right)^2 + O(1/N). \tag{4.3}$$

In short-range models, including spin glasses, however, there is no simple relationship between the two. For example, the uniform spin configuration and that with a single domain wall running along the equator (generated, e.g., using Dobrushin boundary conditions on the ferromagnet at zero temperature in dimensions greater than or equal to two) have an edge overlap of one and a spin overlap of zero.

We emphasize, however, that the edge overlap is only one interesting quantity providing information (in this case, density) about the interface. We will argue that by itself it does not provide sufficient information to distinguish among various interesting pictures of the spin glass phase. Other quantities, in particular the energy scaling of the interface, are also required for a useful description to emerge.

We can now list several reasons why the density, energy, and possibly other variables associated with interfaces between states constitute a useful set of global spin glass variables.

(0) (Preliminary technical point.) *The quantities being studied can be clearly defined.* This is accomplished through the metastate (see Sect. 5) and its natural extensions. As discussed elsewhere, a probability measure on low-energy interfaces can be generated and studied through the periodic boundary condition uniform perturbation metastate [39], while one on higher-energy interfaces can be constructed via a modification of that used in constructing the *excitation metastate* in [36].

(1) *The quantities proposed are truly global; i.e., there is no need to use a "window."* The reason for this is presented in the next point.

(2) *The edge overlap, computed in the entire volume, can provide unambiguous information about pure state multiplicity.* This is in contrast to the spin overlap function. A rigorous formulation and proof of this statement (in the case of zero temperature) was provided in [18]. (The results can be extended to nonzero temperatures by "pruning" from the interfaces small thermally induced droplets [18, 39].) Informally stated, the theorem presented in that paper stated that if edge overlaps were space-filling (that is, their density did not tend to zero with L), then there must be multiple pure state pairs (e.g., in the appropriate periodic b.c. metastate). Otherwise, there is only a single pair.

It needs to be noted here that this result is restricted to boundary conditions chosen *independently* of the couplings (which is always the case in numerical simulations and theoretical computations). It is conceivable that appropriately chosen coupling-*dependent* boundary conditions can generate "interface states," separated by domain walls of vanishing density, as occur below the roughening temperature in Ising ferromagnets; but no procedure for constructing such boundary conditions has yet been found.

(3) *Interface structure also provides information on thermodynamically relevant nonpure state structure, in particular, the distribution and energies of excitations.* This is a potentially important use of the procedures of Krzakala–Martin [40] and Palassini–Young [41], and is described in more detail in [39].

(4) *The scaling of the edge overlap with energy allows one to distinguish between different scenarios of the low-temperature phase at zero temperature.* At zero temperature, the spin overlap function in any volume is identical in the scaling droplet [34, 42–47], chaotic pairs [4, 10, 15–18, 32], and RSB scenarios [1, 2, 6–9, 35] (see the companion paper [28] for a detailed description of these three pictures), but the interface density and energy, when used together, can distinguish among all of these pictures (as well as the "KM/PY" scenario of Krzakala–Martin [40] and Palassini–Young [41]). This is summarized in Fig. 1.

(5) *The interface properties discussed here provide a means of comparing behavior of SK- and short-range spin glasses.* That is, use of interface properties allows comparison without requiring recourse to pure state notions, which as discussed above are poorly defined in the SK model. So, while they provide much (although not all – see below) useful information about

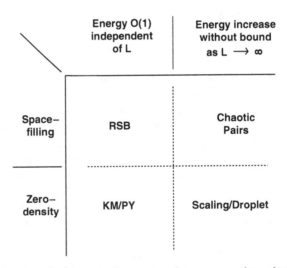

Fig. 1. (Adapted from [48].) Table illustrating the correspondence between interface properties and different scenarios for the structure of the low-temperature phase in short-range spin glasses

pure state structure in short-range models, they are also well-defined in SK models and allow for direct comparison of the two.

It should always be kept in mind, however, that ultimately the interface variables we have been discussing discard a significant amount of important information on local correlations; this is to be expected from any global variable. These variables should therefore be viewed as providing *additional* useful information to the usual local (i.e., thermodynamic) variables; it is dangerous to view them as a replacement for those variables. Consequently, even while providing the arguments in this section, we emphasize that interface – or any other global–variables can play at best a secondary role in describing the statistical mechanics of finite-range models, where well-defined state-based quantities already exist. It is only in the SK model, where such quantities are mostly absent, that global quantities play a more primary – perhaps the only – role.

Throughout much of this section we referred to useful properties of a thermodynamic object we have called the metastate. We conclude by defining this object, which in turn enables us to return to the observation made at the beginning of this note.

5 What are Metastates, Anyway?

We are interested in the thermodynamic behavior of large, finite systems, containing on the order of 10^{23} interacting degrees of freedom. Corresponding infinite volume limits serve two purposes. Mathematically, they enable precise definitions of quantities corresponding to physical variables; physically, they allow one to approximate large finite systems (usually when surface effects can be ignored compared to bulk phenomena). They allow a deep conceptual understanding of important physical phenomena; probably the best-known example is understanding phase transitions in terms of singularities or discontinuities of thermodynamic functions. It is of course a fact that such phenomena correspond to true mathematical singularities only in strictly infinite systems, while common sense dictates that phase transitions in physical systems are quite real, and involve behavior as singular or discontinuous as can be found anywhere in the physical world. That is precisely why infinite volume limits are properly regarded as convenient and useful mathematical descriptions of large finite systems.

When dealing with the usual sort of *global* variable such as energy, we see no serious issues arising in disordered systems, nor do we expect that there is any conceptual divergence from homogeneous systems. Thus, for example, the calculation of the SK free energy per spin in the Parisi solution relies completely on taking the $N \to \infty$ limit (see, e.g., [28]). Any difficulties involved are really of a technical, rather than a conceptual, nature, and they have finally been rigorously resolved [49, 50].

It is when dealing with *local* variables that serious problems arise, and here the behavior of models with quenched disorder seems to diverge dramatically from that of homogeneous ones. In almost all well-known homogeneous models, such as uniform ferromagnets, there is simple convergence, as volume goes to infinity, to a single Gibbs state at high temperature; or to several, related via well-understood symmetry transformations, at low temperature. There the nature of the finite-volume approximation to the infinite-volume limit is conceptually clear and no difficulties arise.

But what if – as has often been conjectured for finite-range spin glasses – there are many pure states and they are not simply related to each other by symmetry transformations? Mathematically, this means that if one looks at a local variable, such as a single-spin or two-spin correlation function, there exist many possible (subsequence) limits. When this happens, it is not even immediately clear how to obtain a well-defined infinite-volume limit. This is largely a consequence of *chaotic size dependence*, first demonstrated in [32] as an unavoidable signature of many states. Briefly put, local variables, such as correlations, will vary chaotically and unpredictably as volume size changes (with, say, periodic boundary conditions on each volume). In fact, this chaotic size dependence was proposed by us as a test of the presence of many pure state pairs. Consequently, convergence of *states* to a thermodynamic limit is no longer as simple as just taking a sequence of volumes of arithmetically, say, or geometrically increasing lengths – a process that works fine for homogeneous systems.

It turns out that such limits *can* be defined, but it takes a little work [10, 14]. However, the existence of limiting thermodynamic *states* turns out not to be the essential problem. As noted, such states do exist and are well-defined; but they turn out not to contain the information needed to fully understand the system *in large finite volumes*. So now there does, at first glance, seem to be a conflict between the thermodynamic limit and behavior in large finite volumes. But such a conclusion is premature – with more work, not only can the two be reconciled again, but an entirely new set of concepts and insights arises.

So we turn to the question: is it even possible in such systems to describe the nature of large finite volume systems via a single infinite volume object, and if so, how? The answer to the first question is yes [15], and to the second is: by an infinite volume object that captures the nature of the behavior of the finite systems as volume increases – i.e., by the *metastate* [4,10,15–17,51] describing the empirical distribution of local variables as volume increases without bound.

Such behavior in L is analogous to chaotic behavior in time t along the orbit of a chaotic dynamical system. In each case the behavior is deterministic but effectively unpredictable. Consequently, it can be modeled via random sampling from some distribution κ on the space of states. In the case of dynamical systems, one can in principle reconstruct κ by keeping a record of the proportion of time the particle spends in each coarse-grained region of state space. Similarly, one can prove [10, 15] that for inhomogeneous systems like spin glasses, a similar distribution exists: even in the presence of chaotic size dependence, the *fraction* of volumes in which a given thermodynamic state Γ appears, converges (at least along a sparse sequence of volumes). By saying that a thermodynamic state Γ (which is an infinite-volume quantity) "appears" within a finite volume Λ_L, we mean the following: within a window deep inside the volume, correlation functions computed using the finite-volume Gibbs state ρ_L are the same as those computed using Γ (with negligibly small deviations).

Hence, the metastate allows one to reconstruct the behavior of large finite volumes from an infinite-volume object, which contains far more information than any mixed thermodynamic state, such as those often used as a starting point in RSB analyses. Because technical details have been provided in several other places [10, 15], we do not repeat them here. We do however refer the reader to [28] where we examine more closely the nature of the low-temperature spin glass phase, from the point of view of metastates.

Acknowledgments

This research was partially supported by NSF Grants DMS-01-02587 (CMN) and DMS-01-02541 (DLS).

References

1. E. Marinari, G. Parisi, and J.J. Ruiz-Lorenzo, in *Spin Glasses and Random Fields*, edited by A.P. Young (World Scientific, Singapore, 1997), pp. 59–98
2. E. Marinari, G. Parisi, F. Ricci-Tersenghi, J.J. Ruiz-Lorenzo, and F. Zuliani, *J. Stat. Phys.* **98**, 973 (2000)
3. H.O. Georgii, *Gibbs Measures and Phase Transitions* (de Gruyter Studies in Mathematics, Berlin, 1988)
4. C.M. Newman and D.L. Stein, *J. Phys.: Cond. Matter* **15**, R1319 (2003)
5. D. Sherrington and S. Kirkpatrick, *Phys. Rev. Lett.* **35**, 1792 (1975)
6. G. Parisi, *Phys. Rev. Lett.* **43**, 1754 (1979)
7. G. Parisi, *Phys. Rev. Lett.* **50**, 1946 (1983)
8. M. Mézard, G. Parisi, N. Sourlas, G. Toulouse, and M. Virasoro, *Phys. Rev. Lett.* **52**, 1156 (1984)
9. M. Mézard, G. Parisi, N. Sourlas, G. Toulouse, and M. Virasoro, *J. Phys. (Paris)* **45**, 843 (1984)
10. C.M. Newman and D.L. Stein, in *Mathematics of Spin Glasses and Neural Networks*, ed. A. Bovier and P. Picco (Birkhäuser, Boston, 1997), pp. 243–287
11. C.M. Newman and D.L. Stein, *Phys. Rev. Lett.* **91**, 197205 (2003)
12. M. Talagrand, *Spin Glasses: A Challenge for Mathematicians: Cavity and Mean Field Models* (Springer, Berlin Heidelberg New York, 2003)
13. R.L. Dobrushin, *Theor. Prob. Appl.* **17**, 582 (1972)
14. C.M. Newman and D.L. Stein, *Phys. Rev. Lett.* **76**, 515 (1996)
15. C.M. Newman and D.L. Stein, *Phys. Rev. Lett.* **76**, 4821 (1996)

16. C.M. Newman and D.L. Stein, *Phys. Rev. E* **55**, 5194 (1997)
17. C.M. Newman and D.L. Stein, *Phys. Rev. E* **57**, 1356 (1998)
18. C.M. Newman and D.L. Stein, *J. Stat. Phys.* **106**, 213 (2002)
19. S. Edwards and P.W. Anderson, *J. Phys. F* **5**, 965 (1975)
20. D. Ruelle, *Statistical Mechanics* (Benjamin, New York, 1969)
21. O. Lanford, in *Statistical Mechanics and Mathematical Problems, Lecture Notes in Physics, v. 20* (Springer, Berlin Heidelberg New York, 1973), pp. 1–135
22. B. Simon, *The Statistical Mechanics of Lattice Gases v. 1*, (Princeton University Press, Princeton, 1993)
23. J. Slawny, in C. Domb and J.L. Lebowitz (eds.), *Phase Transitions and Critical Phenomena v. 11* (Academic, London, 1986), pp. 128–205
24. R.L. Dobrushin and S.B. Shlosman, in S.P. Novikov (ed.), *Soviet Scientific Reviews C. Math. Phys., v. 5* (Harwood Academic, New York, 1985), pp. 53–196
25. A.C.D. van Enter and J.L. van Hemmen, *Phys. Rev. A* **29**, 355 (1984)
26. K. Binder and A.P. Young, *Rev. Mod. Phys.* **58**, 801 (1986)
27. M. Mézard, G. Parisi, and M.A. Virasoro, *Spin Glass Theory and Beyond* (World Scientific, Singapore, 1987)
28. C.M. Newman and D.L. Stein, Short-Range Spin Glasses: Results and Speculations, to appear in 2004 Ascona workshop proceedings
29. A. Bovier and V. Gayrard, in *Mathematics of Spin Glasses and Neural Networks*, ed. A. Bovier and P. Picco (Birkhäuser, Boston, 1997), pp. 3–89
30. A. Bovier, V. Gayrard, and P. Picco, in *Mathematics of Spin Glasses and Neural Networks*, ed. A. Bovier and P. Picco (Birkhäuser, Boston, 1997), pp. 187–241
31. C. Külske, in *Mathematics of Spin Glasses and Neural Networks*, ed. A. Bovier and P. Picco (Birkhäuser, Boston, 1997), pp. 151–160
32. C.M. Newman and D.L. Stein, *Phys. Rev. B* **46**, 973 (1992)
33. A. Bovier, private communication
34. D.A. Huse and D.S. Fisher, *J. Phys. A* **20**, L997 (1987)
35. S. Caracciolo, G. Parisi, S. Patarnello, and N. Sourlas, *J. Phys. (Paris)* **51**, 1877 (1990)
36. C.M. Newman and D.L. Stein, *Phys. Rev. Lett.* **84**, 3966 (2000)
37. E. Marinari and G. Parisi, *Phys. Rev. B* **62**, 11677 (2000)
38. E. Marinari and G. Parisi, *Phys. Rev. Lett.* **86**, 3887 (2001)
39. C.M. Newman and D.L. Stein, *Phys. Rev. Lett.* **87**, 077201 (2001)
40. F. Krzakala and O.C. Martin, *Phys. Rev. Lett.* **85**, 3013 (2000)
41. M. Palassini and A.P. Young, *Phys. Rev. Lett.* **85**, 3017 (2000)
42. W.L. McMillan, *J. Phys. C* **17**, 3179 (1984)
43. A.J. Bray and M.A. Moore, *Phys. Rev. B* **31**, 631 (1985)
44. A.J. Bray and M.A. Moore, *Phys. Rev. Lett.* **58**, 57 (1987)
45. D.S. Fisher and D.A. Huse, *Phys. Rev. Lett.* **56**, 1601 (1986)
46. D.S. Fisher and D.A. Huse, *J. Phys. A* **20**, L1005 (1987)
47. D.S. Fisher and D.A. Huse, *Phys. Rev. B* **38**, 386 (1988)
48. C.M. Newman and D.L. Stein, *Annales Henri Poincaré* **4**, Suppl. 1, S497 (2003)
49. F. Guerra, *Commun. Math. Phys.* **233**, 1 (2003)
50. M. Talagrand, *Ann. Math.* **163**, 221 (2006)
51. M. Aizenman and J. Wehr, *Commun. Math. Phys.* **130**, 489 (1990)

Short-Range Spin Glasses: Results and Speculations

Charles M. Newman and Daniel L. Stein

Courant Institute of Mathematical Sciences, New York University
251 Mercer Street, New York, NY 10012, USA
e-mail: newman@cims.nyu.edu
and

Departments of Physics and Mathematics, University of Arizona
1118 E. 4th Street, Po Box 210081, Tucson, AZ 85721, USA
e-mail: dls@physics.arizona.edu

Summary. This paper is divided into two parts. The first part concerns several standard scenarios for how short-range spin glasses might behave at low temperature. Earlier theorems of the authors are reviewed, and some new results presented, including a proof that, in a thermodynamic system exhibiting infinitely many pure states and with the property (such as in replica-symmetry-breaking scenarios) that mixtures of these states manifest themselves in large finite volumes, there must be an *uncountable* infinity of states.

In the second part of the paper, we offer some conjectures and speculations on possible unusual scenarios for the low-temperature phase of finite-range spin glasses in various dimensions. We include a discussion of the possibility of a phase transition *without* broken spin–flip symmetry, and provide an argument suggesting that in low dimensions such a possibility may occur. The argument is based on a new proof of Fortuin–Kasteleyn random cluster percolation at nonzero temperatures in dimensions as low as two. A second speculation considers the possibility, in analogy to certain phenomena in Anderson localization theory, of a much stronger type of chaotic temperature dependence than has previously been discussed: one in which the actual state space structure, and not just the correlations, vary chaotically with temperature.

Key words: Spin glass, Edwards–Anderson model, Sherrington–Kirkpatrick model, Replica symmetry breaking, Mean–field theory, Pure states, Metastates, Interface, Fortuin–Kasteleyn, Random cluster percolation, Anderson localization

1 Introduction

In this paper we consider Ising spin glass models, in particular the infinite-range Sherrington–Kirkpatrick (SK) model [1] and the nearest neighbor Edwards–Anderson (EA) model [2] on $\mathbf{Z}^d$. The SK Hamiltonian (for N spins) is

$$\mathcal{H}_N = -\left(1/\sqrt{N}\right) \sum_{1 \le i < j \le N} J_{ij}\sigma_i\sigma_j \tag{1.1}$$

where the couplings J_{ij} are independent, identically distributed random variables chosen, e.g., from a Gaussian distribution with zero mean and variance one.

In a series of papers, Parisi and collaborators [3–6] proposed, and worked out the consequences of, an extraordinary *ansatz* for the nature of this phase. Following the mathematical procedures underlying the solution, it came to be known as *replica symmetry breaking* (RSB). The starting point of the Parisi solution was that the low-temperature spin glass phase comprised not just a single spin-reversed pair of states, but rather "infinitely many pure thermodynamic states" [4], not related by any simple symmetry transformations.

What is primarily of interest is the *distribution* of the overlaps between two SK "pure states" [7] α and β, defined as

$$q_{\alpha\beta} = \frac{1}{N} \sum_{i=1}^{N} \langle\sigma_i\rangle_\alpha \langle\sigma_i\rangle_\beta , \tag{1.2}$$

where $\langle\cdot\rangle_\alpha$ is a thermal average in pure state α, and dependence on the coupling realization $\mathcal{J}$ and temperature T has been suppressed. The overlap distribution is constructed by choosing, at fixed N and T, two of the pure states α and β present with probabilities W_α and W_β in the Gibbs state. The probability that their overlap lies between q and $q + dq$ is then given by the quantity $P_{\mathcal{J}}(q)dq$, where

$$P_{\mathcal{J}}(q) = \sum_\alpha \sum_\beta W_{\mathcal{J}}^\alpha W_{\mathcal{J}}^\beta \delta(q - q_{\alpha\beta}) . \tag{1.3}$$

As before, we suppress the dependence on T and N for ease of notation. The average $P(q)$ of $P_{\mathcal{J}}(q)$ over the disorder distribution is commonly referred to as the *Parisi overlap distribution*, and serves as an order parameter for the SK model.

The EA Hamiltonian in zero external field is given by

$$\mathcal{H} = - \sum_{\substack{x,y \\ |x-y|=1}} J_{xy}\sigma_x\sigma_y, \tag{1.4}$$

where the nearest-neighbor couplings J_{xy} are defined in exactly the same way as the J_{ij} in the SK Hamiltonian (1.1). In this model, and unlike in the SK model, thermodynamic pure, mixed, and ground states are standard, well-defined (see, e.g., [8,9]) objects, constructed according to well-established prescriptions of statistical mechanics [10–16].

Now suppose – as has often been conjectured for finite-range spin glasses [17, 18] – that there are many pure states not simply related to each other by symmetry transformations. If this is the case, it was shown in [19] that local variables, such as correlations, will vary chaotically and unpredictably as volume size changes (with, say, periodic boundary conditions on each volume). Is it then even possible to describe the nature of large finite volume systems via a single infinite volume object, and if so, how? It turns out that an object, called the *metastate* [20] by the authors, can be constructed to capture the nature of the behavior inside finite systems as volume increases. The metastate is a useful tool that describes the empirical distribution of local variables as volume increases without bound, and is described in more detail in the companion paper [21] and in [8, 20, 22–25]. The seemingly chaotic behavior with increasing volume is modeled by random sampling from the metastate, regarded as a probability measure on the space of Gibbs states. Using the metastate approach, we now examine more closely the nature of the low-temperature spin glass phase.

2 The Trinity of Scenarios

There are three scenarios that have received the major share of attention in the current literature. Here we refer only to those dealing with the *pure state* structure of the low-temperature spin glass phase. Other pictures have also received a great deal of attention, particularly the excited state scenario of Krzakala–Martin [26] and Palassini–Young [27]. Proving or disproving these or related pictures is important for achieving a thorough understanding of the low-temperature physics of spin glasses. However, because it can be shown [28, 29] that these pictures describe excitations that do *not* alter pure state structure, their presence or absence can be logically incorporated into any of the three pictures that we are about to describe (although on heuristic grounds they appear incompatible with the RSB picture).

Low-temperature pictures of spin glass long-range order and broken symmetry start with an assumption about the number of pure states $\mathcal{N}(\beta, d)$ (which is the same for almost every coupling realization $\mathcal{J}$ – see, e.g., [22]). They assume also a putative critical (inverse) temperature $\beta_c(d) < \infty$ separating a paramagnetic phase for $\beta < \beta_c$ from a spin glass phase for $\beta > \beta_c$. Although there is good numerical [17, 30–32] and some analytical [33, 34] evidence that above some lower critical dimension d_c^l there does exist such a finite $\beta_c(d)$, there is as yet no proof or even a strong physical argument supporting such a conjecture. Moreover, there is no logical reason why there cannot be *two* or more phase transitions in some dimensions. However, we will not attempt to enumerate all possibilities here; we will confine ourselves to the most likely scenarios. We defer to Sect. 4 a consideration (mostly for the fun of it) of some of the more outlandish sounding possibilities.

The actual value of $\mathcal{N}(\beta, d)$ is not rigorously known at large β for any $d \geq 2$. There does exist a rigorous argument [35] supporting – but not completely proving – the conjecture that $\mathcal{N}(\infty, 2) = 2$, and a heuristic argument [36] supporting the conjecture that $\mathcal{N}(\beta < \infty, 2 < d < 8) \leq 2$. It is generally assumed (but also not proved) that spin–flip symmetry is broken for $\beta_c < \beta < \infty$, so that pure states come in global spin–flip reversed pairs.

Given all these assumptions, pure state scenarios generally assume either a single pair of pure states or infinitely many. Again, there is at this time no argument proving that one cannot have, say, $\mathcal{N}(\beta, d) = 20$; it is just difficult to imagine why such a scenario should occur. There is, however, a reasonably strong heuristic argument [24] indicating that if $\mathcal{N}(\beta, d) = \infty$, it must be an *uncountable* infinity (we will prove this in Sect. 3 for the RSB picture).

While an assumption about $\mathcal{N}(\beta, d)$ is the *starting point* for each of the three pictures we now describe, they are each much more than a simple assertion about the number of pure states. In particular, there are potentially many two-state or many-state pictures that correspond to *none* of these three scenarios. Here the only aspect of the three on which we will focus is the relationship among the pure states.

2.1 Heuristic Description

The scenarios are, in order of increasing complexity, the scaling/droplet (SD) picture [37–43], the chaotic pairs (CP) picture [8,9,19,20,23,24], and the RSB picture [3–6,18]. The first is a 2-state scenario, while the second and third are many-state scenarios. The pure state structure in these pictures is normally described through the Parisi overlap function $P(q)$ [18]. This function needs to be used with great care in describing pure state structure in short-range models; see [8,24,41] for a discussion of some of the pitfalls and problems that can occur in its applications.

In Sect. 2.2 we provide precise definitions of the pure state structure of these pictures; we limit ourselves here to brief heuristic descriptions. As already noted, scaling/droplet is a two-state picture. The overlap distribution is therefore simply a pair of δ-functions at $q = \pm q_{EA}$, where q_{EA} is the Edwards–Anderson order parameter [2], regardless of whether replicas and overlaps are taken before or after the thermodynamic limit is taken, as shown in Fig. 1 (see especially [8,9,22] for a more detailed discussion of the difference between the two procedures).

The CP picture considers the possibility of an infinite number of pure state pairs, but with a trivial overlap structure: all large, finite volumes would display an overlap structure equivalent to the SD picture because only a single pair of pure states appears in each of these volumes. The actual pure state pair, however, varies chaotically as volume changes because different volumes select different members from the infinite ensemble of such pairs. The overlap

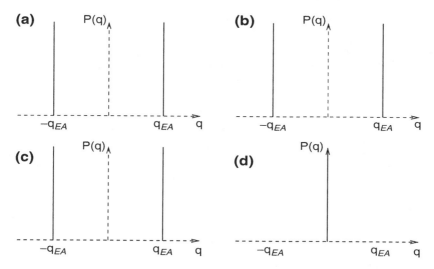

Fig. 1. (From [9].) The spin overlap function $P(q)$ at $T < T_c$ for: **(a)** a two-state picture when replicas are taken *before* the thermodynamic limit; **(b)** the many-state chaotic pairs picture when replicas are taken before the thermodynamic limit; **(c)** a two-state picture when replicas are taken *after* the thermodynamic limit; **(d)** the many-state chaotic pairs picture when replicas are taken after the thermodynamic limit (conjectured)

distribution for the infinite volume in the CP picture, constructed by choosing replicas and taking their overlaps *after* going to the thermodynamic limit, would then presumably be a simple δ-function at zero overlap (cf. Fig. 1).

There are actually two RSB pictures, which we have called "standard" and "nonstandard" SK [20, 23, 24, 44]. The first chooses replicas *after* taking the thermodynamic limit, and the second before. Both have nontrivial overlap structures, which will be described in Sect. 2.2.

Before turning to that, we need to say a few words about the dimension dependence of the three pictures. The only one of these with specific predictions is the SD picture, which asserts that in *every* finite dimension where spin–flip symmetry is broken, there is only a single pure state pair. The CP picture does not make any corresponding claim; it merely asserts that *if* $\mathcal{N}(\beta, d) = \infty$ at some (β, d), *then* the overlap structure must be trivial in the manner specified above. If the RSB picture correctly describes the EA model at $d = \infty$, then both the SD and CP models agree that the upper critical dimension $d_c^u = \infty$ for the EA model.

The RSB picture is slightly less vague about its dimension-dependence; it does apparently assume that there exists a strictly finite d_c^u for the EA model, and that replica symmetry is broken in the nontrivial manner it specifies for all $d \geq d_c^u$. The precise value of d_c^u within this picture remains uncertain, but it seems to be higher than two and less than or equal to three [45, 46].

2.2 Description via the Metastate

We now turn to a precise description of the three competing pictures, which is greatly facilitated by using the metastate.

In SD, there is only a single pair of pure states, and these are the same in every large volume; the overlap distribution function $P_{\mathcal{J}}^L(q)$ in a volume Λ_L therefore simply approximates a sum of two δ-functions at $\pm q_{EA}$, as shown in Fig. 1 a. In the CP picture, each finite-volume Gibbs state $\rho_{\mathcal{J}}^L$ will still be approximately a mixture of a *single* pair of spin–flip-related pure states, *but now the pure state pair will vary chaotically with L*. Then for each Λ_L, $P_{\mathcal{J}}^L(q)$ will again approximate a sum of two δ-functions at $\pm q_{EA}$.

So chaotic pairs resembles the scaling/droplet picture in finite volumes, but has a very different thermodynamic structure. It is a many-state picture, but differs from the RSB picture (see below) in that only a *single* pair of spin-reversed pure states $\rho_{\mathcal{J}}^{\alpha_L}$, $\rho_{\mathcal{J}}^{\overline{\alpha_L}}$, appears in a large volume Λ_L with symmetric boundary conditions, such as periodic. In other words, for large L, one finds that

$$\rho_{\mathcal{J}}^{(L)} \approx \frac{1}{2}\rho_{\mathcal{J}}^{\alpha_L} + \frac{1}{2}\rho_{\mathcal{J}}^{\overline{\alpha_L}}, \tag{2.1}$$

and the pure state pair (of the infinitely many present) appearing in a particular finite volume Λ_L depends chaotically on L. That is, the periodic b.c. metastate is dispersed over many Γ's (in fact, an uncountable infinity, as we shall see below; this also occurs in the RSB picture), but (unlike in RSB) each Γ is a *trivial* mixture of the form $\Gamma = \Gamma^\alpha = \frac{1}{2}\rho_{\mathcal{J}}^\alpha + \frac{1}{2}\rho_{\mathcal{J}}^{\overline{\alpha}}$. The overlap distribution for each Γ is the same: $P_\Gamma = \frac{1}{2}\delta(q - q_{EA}) + \frac{1}{2}\delta(q + q_{EA})$. It is interesting to note that there is a spin glass model (the "highly disordered model" [47–49]) that appears to display just this behavior in its ground state structure above eight dimensions.

We now turn to the standard and nonstandard mean-field-like scenarios. The standard SK picture is perhaps most concisely described in the introduction of [50]. It requires a Gibbs equilibrium measure $\rho_{\mathcal{J}}(\sigma)$ which is decomposable into many pure states $\rho_{\mathcal{J}}^\alpha(\sigma)$:

$$\rho_{\mathcal{J}}(\sigma) = \sum_\alpha W_{\mathcal{J}}^\alpha \rho_{\mathcal{J}}^\alpha(\sigma). \tag{2.2}$$

In this picture replicas are taken *after* going to the thermodynamic limit. That is, one chooses σ and σ' from the product distribution $\rho_{\mathcal{J}}(\sigma)\rho_{\mathcal{J}}(\sigma')$, and then the overlap can be defined as

$$Q = \lim_{L \to \infty} |\Lambda_L|^{-1} \sum_{x \in \Lambda_L} \sigma_x \sigma_x', \tag{2.3}$$

where $|\Lambda_L|$ is the volume of the cube Λ_L.

Suppose that σ is drawn from pure state $\rho_{\mathcal{J}}^\alpha$ and σ' from $\rho_{\mathcal{J}}^\gamma$. Then (7) equals its thermal mean [44]

$$q_{\mathcal{J}}^{\alpha\gamma} = \lim_{L\to\infty} |\Lambda_L|^{-1} \sum_{x\in\Lambda_L} \langle\sigma_x\rangle_\alpha \langle\sigma_x\rangle_\gamma, \tag{2.4}$$

and so the overlap distribution $P_{\mathcal{J}}(q)$ is given by

$$P_{\mathcal{J}}(q) = \sum_{\alpha,\gamma} W_{\mathcal{J}}^\alpha W_{\mathcal{J}}^\gamma \delta(q - q_{\mathcal{J}}^{\alpha\gamma}). \tag{2.5}$$

According to this picture, the $W_{\mathcal{J}}^\alpha$'s and $q_{\mathcal{J}}^{\alpha\gamma}$'s are non-self averaging quantities, except when $\alpha = \gamma$ or its global flip, where $q_{\mathcal{J}}^{\alpha\gamma} = \pm q_{EA}$. The average $P(q)$ of $P_{\mathcal{J}}(q)$ over the disorder distribution ν of the couplings is then a mixture of two delta-function components at $\pm q_{EA}$ and a continuous part between them.

There is a technical problem (caused by chaotic size dependence [19]) in the construction of $\rho_{\mathcal{J}}(\sigma)$ that can be overcome by using the periodic b.c. metastate $\kappa_{\mathcal{J}}^{\text{PBC}}$ (in fact, any coupling-independent metastate would do). One can construct [44] a state $\rho_{\mathcal{J}}(\sigma)$ which is the *average* over the metastate:

$$\rho_{\mathcal{J}}(\sigma) = \int \Gamma(\sigma)\kappa_{\mathcal{J}}(\Gamma)\, d\Gamma. \tag{2.6}$$

One can also think of this $\rho_{\mathcal{J}}$ as the average thermodynamic state, $N^{-1}(\rho_{\mathcal{J}}^{(L_1)} + \rho_{\mathcal{J}}^{(L_2)} + \ldots, \rho_{\mathcal{J}}^{(L_N)})$, in the limit $N\to\infty$. It can be proved [22, 25] that $\rho_{\mathcal{J}}(\sigma)$ is a Gibbs state.

Numerically one constructs overlaps without constructing Gibbs states at all. Such a construction (similar to that above) is described in [44], and leads ultimately to the same conclusion.

But this picture can be rigorously ruled out for the EA model [44], as will be shown in Sect. 3. So the most natural (and usual) interpretation of a mean-field-like picture cannot be applied to short-range spin glasses. The question then becomes: are there alternative, less straightforward interpretations? To address this question, we constructed in [20] (see also [8, 9, 22–24]) an alternative mean-field-like picture, the "nonstandard SK model," which we hereafter refer to simply as the RSB picture. This picture clarifies as well how broken replica symmetry should be interpreted for the SK model.

Consider again the PBC metastate, although, as always, almost any other coupling-independent metastate will suffice. The RSB picture assumes that in each volume Λ_{L_i}, the finite-volume Gibbs state $\rho_{\mathcal{J},L_i}$ is well approximated deep in the interior by a mixed thermodynamic state $\Gamma^{(L_i)}$, decomposable into many pure states $\rho_{\alpha_{L_i}}$:

$$\Gamma^{(L_i)} = \sum_{\alpha_{L_i}} W_{\Gamma^{(L_i)}}^{\alpha_{L_i}} \rho_{\alpha_{L_i}} \tag{2.7}$$

where explicit dependence on $\mathcal{J}$ is suppressed.

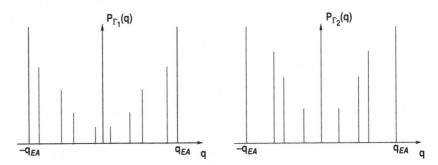

Fig. 2. (From [8].) The overlap distribution, at fixed $\mathcal{J}$, in two different volumes Λ_1 and Λ_2 in the nonstandard SK picture

As in the chaotic pairs picture, the mixed states $\Gamma^{(L_i)}$ change in some "chaotic" fashion with L_i. Furthermore, each mixed state $\Gamma^{(L_i)}$ is presumed to have a nontrivial overlap distribution

$$P_{\Gamma^{(L_i)}} = \sum_{\alpha_{L_i}, \beta_{L_i}} W^{\alpha_{L_i}}_{\Gamma^{(L_i)}} W^{\beta_{L_i}}_{\Gamma^{(L_i)}} \delta(q - q_{\alpha_{L_i} \beta_{L_i}}) \qquad (2.8)$$

of the form shown in Fig. 2. Moreover, the distances among any three pure states *within a particular* Γ are assumed to be ultrametric [18].

We conclude with a brief note about zero temperature. In each Λ_L (with periodic b.c.'s) there can only be a single ground state pair at $T = 0$ (because we are considering Gaussian couplings rather than a $\pm J$ model). If scaling/droplet holds, then this pair is the same for all large L; if infinitely many ground state pairs exist, then the pair changes chaotically with L. In this respect the behavior of CP is the same at both zero and nonzero temperatures (below T_c), and the same can be said for SD. So the overlap functions of CP and RSB differ only at positive temperature: the mean-field RSB picture at $T > 0$ has the $\Gamma^{(L)}$ in each volume exhibiting a nontrivial mixture of pure state pairs as in (2.7), while in chaotic pairs the $\Gamma^{(L)}$ appearing in any Λ_L consists of a single pure state pair, as in (2.1). There remains, however, an important difference between the CP and RSB pictures at zero temperature, in that the space-filling interfaces between ground states should have energies that scale differently; see [8] and the companion paper in this volume [21] for a more detailed discussion.

3 Replica Symmetry Breaking for Short-Range Models

We begin by eliminating the standard SK picture from further consideration; the detailed rigorous proof appears in [44]. Using the torus-translation symmetry of the periodic b.c.'s, one can show that the Gibbs state $\rho_{\mathcal{J}}(\sigma)$ is translation-*covariant*; that is, $\rho_{\mathcal{J}^a}(\sigma) = \rho_{\mathcal{J}}(\sigma^{-a})$, or in terms of correlations,

$\langle\sigma_x\rangle_{\mathcal{J}^a} = \langle\sigma_{x-a}\rangle_{\mathcal{J}}$, and similarly for n-point correlations. Translation covariance of $\rho_{\mathcal{J}}$ immediately implies, via (2.3)–(2.5), translation invariance of $P_{\mathcal{J}}$. But, given the translation-ergodicity of the underlying disorder distribution ν, *it immediately follows that $P_{\mathcal{J}}(q)$ is self-averaging, and equals its distribution average $P(q)$ for a.e. $\mathcal{J}$.* The impossibility of a non-selfaveraging $P_{\mathcal{J}}(q)$ can also be shown for other coupling-independent b.c.'s, where torus-translation symmetry is absent, using methods described in [35].

This leaves nonstandard SK as a maximally allowed mean-field-type picture. Before discussing its viability, we explain why neither CP nor nonstandard SK remains viable if modified to have only a *countable* infinity of pure states. We do this first by a heuristic argument valid for both CP and nonstandard SK, and then by a rigorous argument valid only for nonstandard SK. The conclusion of this exercise is that both alternatives to the SD scenario of a single pure state pair require *uncountably many* (i.e., a continuum of) pure state pairs for a single fixed $\mathcal{J}$.

Consider $\rho_{\mathcal{J}}$, the average over the periodic b.c. metastate $\kappa_{\mathcal{J}}$. Changing from periodic boundary conditions to antiperiodic, or to any of the many other "partially antiperiodic" boundary conditions related to periodic ones by a gauge transformation, leaves $\kappa_{\mathcal{J}}$ and consequently $\rho_{\mathcal{J}}$ unchanged [24]. On heuristic grounds, unless $\rho_{\mathcal{J}}$ is of the SD form with only a single pair of pure states, then this lack of dependence on boundary conditions should hold only if $\rho_{\mathcal{J}}$ is a *uniform* mixture over infinitely many pure states $\rho_{\mathcal{J}}^{\alpha}$ – i.e., the relative weights of all $\rho_{\mathcal{J}}^{\alpha}$'s in $\rho_{\mathcal{J}}$ are equal. But that is clearly impossible if the number of $\rho_{\mathcal{J}}^{\alpha}$'s is countably infinite.

In the case of an SK-type picture, where the Γ's appearing in the metastate $\kappa_{\mathcal{J}}$ involve nontrivial weights $W_{\mathcal{J}}^{\alpha}$, one can go further and *prove* that nontrivial $W_{\mathcal{J}}^{\alpha}$'s require uncountably many $\rho_{\mathcal{J}}^{\alpha}$'s to appear in $\kappa_{\mathcal{J}}$. The argument proceeds as follows. If there are only countably many $\rho_{\mathcal{J}}^{\alpha}$'s, then the corresponding weights (say, ordered by magnitude) in $\rho_{\mathcal{J}}$ will be measurable and translation-invariant functions of $\mathcal{J}$. By translation-ergodicity, this means that they are in fact *independent* of $\mathcal{J}$ and so remain the same if finitely many couplings J_{xy} change by amounts ΔJ_{xy}. Moreover, one can consider the finitely many α's corresponding to, say, the largest k weights in $\rho_{\mathcal{J}}$, and then the *distribution* of their weights within the Γ's of the metastate $\kappa_{\mathcal{J}}$ is also independent of $\mathcal{J}$.

On the other hand, this leads to a contradiction, because a Γ in the nonstandard SK metastate (see (2.7)) is a nontrivial mixture of pure states and their weights. By the Aizenman–Wehr transformation [25], a pure state ρ_{α} transforms to a pure state $\rho_{\alpha'}$ under such a finite change $\mathcal{J} \to \mathcal{J} + \Delta \mathcal{J}$, and its weight W_{α} within Γ correspondingly changes according to

$$W_{\alpha} \to W_{\alpha'} = r_{\alpha} W_{\alpha} / \sum_{\gamma} r_{\gamma} W_{\gamma} \qquad (3.1)$$

where

$$r_{\alpha} = \langle\exp(\beta \sum_{\langle xy\rangle} \Delta J_{xy}\sigma_x\sigma_y)\rangle_{\alpha}. \qquad (3.2)$$

For pure states α and δ that are not identical one can find a choice of $\Delta\mathcal{J}$, with all ΔJ_{xy} small, such that $r_\alpha \neq r_\delta$. Then for all Γ's in which $W_\alpha(\Gamma)$ and $W_\delta(\Gamma)$ are nonzero, the ratio $W_{\alpha'}(\Gamma)/W_{\delta'}(\Gamma)$ is changed from $W_\alpha(\Gamma)/W_\delta(\Gamma)$ by the same factor r_α/r_δ, that is close to 1. But this implies that $\mathcal{J} \to \mathcal{J}+\Delta\mathcal{J}$ will change the distribution of the weights within the metastate, leading to a contradiction. We have thus sketched the proof of:

Theorem 3.1. *The PBC metastate $\kappa_\mathcal{J}$ for the EA spin glass cannot assign strictly positive probability to Γ's whose pure state decompositions satisfy both:*
(a) $\Gamma = \sum_\alpha W_\mathcal{J}^\alpha \rho_\mathcal{J}^\alpha$, *with not all $W_\mathcal{J}^\alpha = 1/2$, and*
(b) over all these Γ's only countably many $\rho_\mathcal{J}^\alpha$'s appear.

An immediate consequence of Theorem 3.1, when combined with our earlier arguments, is that the only remaining mean-field-like picture is the nonstandard SK model with a (periodic b.c.) metastate $\kappa_\mathcal{J}$ whose average $\rho_\mathcal{J}$ is supported on an *uncountable* infinity of pure states.

We cannot at this point rule this scenario out rigorously, but can provide a strong heuristic argument, based on a rigorous result, that makes it very unlikely. The rigorous result is the already mentioned *invariance of the meta-state* [24]: two metastates constructed using gauge-related b.c.'s (e.g., periodic and antiperiodic, or any two randomly chosen fixed b.c.'s) are *identical*. This makes it difficult to see how any many-state picture can have a $\rho_\mathcal{J}$ supported on anything other than a uniform distribution of pure states. But if this did occur for a particular $\mathcal{J}$, an Aizenman–Wehr transformation suggests that the uniformity would be destroyed for a finitely different $\mathcal{J}'$. (For detailed arguments, see [24].)

Why does not this argument also rule out chaotic pairs? Because in the chaotic pairs picture, as in scaling/droplet, there are in each $\Gamma^{(L)}$ only two pure states (although in CP the pair depends on L), each with weight $1/2$. All even correlations are the same in any pair of flip-related pure states, so, by (3.1) and (3.2), any change in couplings leaves the weights unchanged.

We conclude that the only viable many-state scenario for short-range Ising spin glasses is the CP picture with an uncountable infinity of pure states. This picture has trivial replica symmetry breaking (cf. Fig. 1) and consequently a very simple overlap structure.

4 Wild Possibilities

In Sect. 2 we presented the most likely scenarios for the pure state structure of the spin glass phase in short-range models. One of these, the RSB picture, although a priori plausible, especially given its relevance to the SK model, was rigorously ruled out in two of its possible versions and excluded heuristically in its final remaining version. However, there are other, more exotic scenarios that might also occur but do not seem to have been considered. Here, for the sake of completeness (and perhaps a little whimsy) we suggest two. In order to keep the list of possibilities relatively constrained, we will assume in all that (a) a thermodynamic phase transition does exist above some lower critical dimension d_c^l, and (b) in a fixed dimension the low-temperature phase does not alternate among different scenarios, such as SD and CP, as temperature is lowered (although the third and most exotic scenario may be regarded as a kind of phase of extreme alternation).

4.1 Phase Transition Without Broken Spin–Flip Symmetry

As mentioned in Sect. 2, despite decades of effort and considerable numerical support [17, 30–32], there remains no proof of a phase transition in EA spin glasses. In this section we provide a proof that shows that a necessary (though not sufficient) condition for broken spin–flip symmetry at $T > 0$ is satisfied in lattices with dimension as low as two (e.g., a triangular lattice). We also discuss a corresponding sufficient condition. Of course, this (minimal) symmetry-breaking is assumed in all scenarios discussed in Sect. 2; but here we speculate on the possibility of a phase transition *without* the appearance of multiple Gibbs states, and discuss why the presence of the necessary condition (for broken spin–flip symmetry) at some (β, d) where the sufficient condition is absent suggests just such a possibility.

Although several approaches are possible, we use here the Fortuin–Kasteleyn random cluster (RC) representation [51, 52] extended to nonferromagnetic models (see, e.g., [53]), which relates the statistical mechanics of Ising (or Potts) models to a dependent percolation problem. We let $\mathbf{E}^d$ denote the set of bonds $\langle x, y \rangle$ in some d-dimensional lattice $\mathbf{L}^d$, each with corresponding coupling J_{xy}. The RC approach introduces parameters $p_{xy} \in [0, 1)$ by:

$$p_{xy} = 1 - \exp[-\beta |J_{xy}|] . \tag{4.1}$$

The RC distribution is a probability measure μ_{RC} on $\{0, 1\}^{\mathbf{E}^d}$, that is, on 0- or 1-valued bond occupation variables n_{xy}. It is one of two marginal distributions (the other being the ordinary Gibbs distribution) of a joint distribution on $\Omega = \{-1, +1\}^{\mathbf{L}^d} \times \{0, 1\}^{\mathbf{E}^d}$ of the spins and bonds together, and is given by (formally, in the infinite system) by

$$\mu_{\mathrm{RC}}(n_{xy}) = Z_{\mathrm{RC}}^{-1} \, 2^{\#(\{n_{xy}\})} \, \mu_{\mathrm{ind}}(\{n_{xy}\}) \, 1_U(\{n_{xy}\}) , \tag{4.2}$$

where Z_{RC} is a normalization constant, $\#(\{n_{xy}\})$ is the number of clusters determined by the realization $\{n_{xy}\}$, $\mu_{\mathrm{ind}}(\{n_{xy}\})$ is the Bernoulli product measure corresponding to independent occupation variables with $\mu_{\mathrm{ind}}(\{n_{xy} = 1\}) = p_{xy}$, and 1_U is the indicator function on the event U in $\{0, 1\}^{\mathbf{E}^d}$ that there exists a choice of the spins σ_x so that $J_{xy} n_{xy} \sigma_x \sigma_y \geq 0$ for all $\langle x, y \rangle$ [54–56]. U is the event that there is no frustration in the occupied bond configuration. Finite-volume versions of the above formulas, with specified boundary conditions, are similarly constructed.

The mapping of this formalism to ferromagnets follows from formulae (that do not hold for nonferromagnets) such as [57]

$$\langle \sigma_x \sigma_y \rangle = \mu_{\mathrm{RC}}(x \leftrightarrow y), \tag{4.3}$$

where $\langle \sigma_x \sigma_y \rangle$ is the usual Gibbs two-point correlation function and μ_{RC} $(x \leftrightarrow y)$ is the RC probability that x and y are in the same cluster. Similarly, with appropriate RC (wired) boundary conditions used for μ_{RC}, one has

$$\langle \sigma_x \rangle_+ = \mu_{\mathrm{RC}}(x \leftrightarrow \infty). \tag{4.4}$$

It follows that, for ferromagnets, a phase transition from a unique (paramagnetic) phase at low β to multiple infinite volume Gibbs distributions at large β is equivalent to a percolation phase transition for the corresponding RC measure.

For spin glasses (or other nonferromagnets) the situation is less straightforward. Now for two sites x and y, (4.3) becomes

$$\langle \sigma_x \sigma_y \rangle = \langle 1_{x \leftrightarrow \infty} \eta(x, y) \rangle_{RC}; \quad \eta(x, y) = \prod_{\langle x'y' \rangle \in \mathcal{C}} \mathrm{sgn}(J_{x'y'}), \tag{4.5}$$

where $\mathcal{C}$ is any path of occupied bonds from x to y. By the definition of U, any two paths $\mathcal{C}$ and $\mathcal{C}'$ in the *same* cluster will satisfy $\prod_{\langle x'y' \rangle \in \mathcal{C}} \mathrm{sgn}(J_{x'y'}) = \prod_{\langle x'y' \rangle \in \mathcal{C}'} \mathrm{sgn}(J_{x'y'})$.

It is no longer evident that RC percolation is sufficient to prove broken spin–flip symmetry. Consider the case of a finite volume Λ_L with fixed boundary conditions, i.e., a specification $\overline{\sigma}_x = \pm 1$ for each $\overline{\sigma}_x \in \partial \Lambda_L$. For the ferromagnet, by first choosing all $\overline{\sigma}_x = +1$ and then all $\overline{\sigma}_x = -1$, one can change the sign of the spin σ_0 at the origin even as $L \to \infty$. That is, boundary conditions infinitely far away affect σ_0, which is a signature of the existence of multiple Gibbs distributions.

But for the EA spin glass, it is not clear whether, even in the presence of RC percolation, there exist any two sets of boundary conditions with the same effect. Although the infinite cluster in any one RC realization is unique, different RC realizations can have different paths from $0 \leftrightarrow \partial \Lambda_L$, leading to different signs for σ_0. So percolation might still allow for $\langle \sigma_0 \rangle \to 0$ as $L \to \infty$.

However, it is easy to see that single RC percolation is at least a *necessary* condition for multiple (symmetry-broken) Gibbs phases. The contribution to the expectation of σ_0 from any finite RC cluster is zero: if a spin configuration σ is consistent with a given RC bond realization within such a cluster, $-\sigma$ is also consistent and equally likely. As a consequence, at a (β, d) where RC percolation does not hold, $\langle \sigma_0 \rangle = 0$ in infinite volume. We note that a slightly stronger version of this argument [56] proves that the transition temperature for an EA $\pm J$ (or other) spin glass, if it exists, is bounded from above by the transition temperature in the corresponding (disordered) ferromagnet.

Is there a modification of this approach that could lead to a proof of multiple Gibbs phases? One such possibility is what might be called *double* RC percolation. Here one expands the sample space Ω to include two independent copies of the bond occupation variables (for a given $\mathcal{J}$ configuration), and defines the variable $r_{xy} = n_{xy}n'_{xy}$, where n_{xy} and n'_{xy} are taken from the two copies. One then replaces percolation of $\{n_{xy}\}$ in the single RC case with percolation of $\{r_{xy}\}$. It is not hard to see that this would be a sufficient condition for the existence of multiple Gibbs phases (and consequently, for a phase transition).

Single RC percolation for the EA $\pm J$ Ising model in $d > 2$ [55] has been proved [53]. Here we sketch the outline of a proof (simpler than the one in [53]) showing that one has single RC percolation *even in* $d = 2$ – e.g., on the triangular lattice. This is already interesting for the following reason: because we are not aware of strong numerical evidence of multiple Gibbs states for this geometry, this may be considered evidence that single RC percolation can occur in spin glasses without broken spin–flip symmetry. But there is an even more interesting potential consequence, which we will discuss below.

The proof uses a standard Fortuin–Kasteleyn–Ginibre (FKG) argument [58] using correlation inequalities. Let $\mathbf{E}_L$ be the edge set confined entirely within a volume Λ_L. It can be shown that there exist probability measures ν_L on $\{0,1\}^{\mathbf{E}_L}$ such that, if $f(\{n\})$ and $g(\{n\})$ are nondecreasing real functions (that is, they do not decrease when any $\{n\} \to \{n'\}$, where every $n_{xy} = 1$ corresponds to $n'_{xy} = 1$, but $n_{xy} = 0$ can correspond to either $n'_{xy} = 0$ or 1), then they are positively correlated:

$$\langle fg \rangle_{\nu_L} \geq \langle f \rangle_{\nu_L} \langle g \rangle_{\nu_L} . \tag{4.6}$$

One measure satisfying this property is the independent product measure [59]. Consequently, the marginal distribution (i.e., averaged over the coupling (signs) in the $\pm J$ model) of satisfied bonds (i.e., using the "satisfaction" variables $\tilde{n}_{xy} = 1$ (or else 0) if $J_{xy}\sigma_x\sigma_y > 0$) at $\beta = 0$, satisfies this property. Operationally, one can think of constructing this set by choosing the spins *first* through independent flips of a fair coin, and *then* choosing the sign of each bond in the same way. One then has independent, density-1/2 bond occupation, which percolates on the triangular lattice [60, 61].

Consider now the satisfaction variables $\tilde{n}_b$ for $\beta > 0$. It is not hard to show that, at any such fixed β, and for any bond $b = \langle xy \rangle$,

$$P(\tilde{n}_b = 1 | \{\tilde{n}_{b'} : b' \neq b\}) \geq 1/2 + \epsilon(\beta) , \tag{4.7}$$

where $\epsilon(\beta)$ is a nonnegative function of β. Equation (4.7) implies that percolation of satisfied bonds at finite temperature dominates (in the FKG sense) percolation at infinite temperature. Since the RC variables n_b are obtained from the satisfaction variables $\tilde{n}_b$ by a slight (for large β) dilution, it follows that at sufficiently low (but nonzero) temperature one has single RC percolation.

It is presumably the case that on the triangular lattice there is no broken spin–flip symmetry, and also only a single Gibbs state, at all nonzero temperatures. But it is worth entertaining the possibility that single, but no double

RC percolation, *does* imply some sort of phase transition *but with a single Gibbs state at all nonzero temperatures.*

Let $\beta_c < \infty$ be the inverse RC percolation transition temperature for the EA model on the triangular lattice. Consider again the expectation of the spin at the origin, $\langle \sigma_0 \rangle$, in a volume Λ_L with plus boundary conditions. For $\beta < \beta_c$, the probability of the site 0 belonging to a cluster reaching the boundary is bounded from above by $c_0(\beta)e^{-c_1(\beta)L}$, where each $c_i(\beta) > 0$ is a finite constant. This would imply that $\langle \sigma_0 \rangle \le c_0(\beta)e^{-c_1(\beta)L}$.

For $\beta > \beta_c$, however, the probability that the origin belongs to a cluster reaching the boundary is bounded away from zero as $L \to \infty$. In order for no phase transition to take place, there must be at fixed β an almost exact cancellation between those RC realizations of "positive" (in the product of the signs of the couplings) and "negative" paths from the origin to the boundary. It is at least conceivable that, while $\sigma_0 \to 0$ as $L \to \infty$, it falls off slower than exponentially – perhaps as a power law in L. This would imply a phase transition from a paramagnetic phase at high temperature to a phase at low temperature with a *unique* Gibbs state, but one where two-point correlations decay as a power law: $\langle \sigma_x \sigma_y \rangle \sim |x - y|^{-\alpha(\beta)}$ as $|x - y| \to \infty$. This would be analogous to the Kosterlitz–Thouless phase transition for $2D$ XY models [62].

4.2 Highly Chaotic Temperature Dependence

We discussed earlier [19] the correspondence between multiple Gibbs state *pairs* and the appearance of chaotic size dependence of correlations in an infinite sequence of volumes with, say, periodic boundary conditions. A possibly related phenomenon is the speculated presence in spin glasses (both infinite- and short-range) of *chaotic temperature dependence* (CTD). Roughly speaking, this refers to an erratic behavior of correlations, upon changing temperature, on lengthscales that diverge as the temperature increment goes to zero. CTD was predicted [39, 43, 63] for the EA spin glass within the SD context; it also seems to be implied by the RSB theory [64–66]. More recent numerical and analytical work (see [67] and references therein) have led to claims that CTD is not present in either the SK or the EA model (see also [68]), although [69] allows for the possibility of a weak effect at large lattice sizes. A perturbative approach [70] observes a very small effect at ninth order. At this time the issue remains unresolved. Its potential presence in spin glasses is interesting, however, and represents a qualitatively new thermodynamic feature of at least some types of disordered systems.

In this section we raise the possibility of a far stronger version of CTD, which we will call *highly* chaotic temperature dependence (HCTD) [36]. Unlike "ordinary" CTD, where correlations behave in a chaotic fashion as temperature changes but the global pure state structure does not, in HCTD the number of pure states in the periodic b.c. metastate does not behave in a continuous or even monotonic fashion as temperature is lowered. This picture departs radically from any other that has appeared so far in the literature.

The HCTD scenario is summarized as follows. As in all other pictures, there exists a deterministic T_c for a.e. $\mathcal{J}$, such that for all $T > T_c$ there is a unique (paramagnetic) Gibbs state. Suppose one were now to choose an arbitrary (nonzero) $T < T_c$. Then with probability one (i.e., for almost every $\mathcal{J}$), there is again a unique infinite volume Gibbs state (though maybe not with exponential decay of truncated correlations). Nevertheless, as temperature is lowered from T_c to 0 for a fixed typical $\mathcal{J}$, there would be a (countably infinite) dense set of temperatures, *depending on* $\mathcal{J}$, with broken symmetric pure phases for that $\mathcal{J}$ at those temperatures.

Now this scenario might be dismissed, with some justification, as the authors' fevered imaginings, but it should be noted that just this sort of phenomenon does happen in other disordered systems. Consider Anderson localization in one dimension, with the random Schrödinger operator [71]

$$\mathcal{H}_\omega = -\nabla^2 + V_0(x) + \lambda V_\omega(x) , \tag{4.8}$$

where $\lambda > 0$, $V_0(x)$ is a bounded potential periodic in x, and

$$V_\omega(x) = \sum_{n=-\infty}^{\infty} \eta_n^\omega U(x - x_n) . \tag{4.9}$$

In (4.9) n runs over the integers, $U(x - x_n)$ is a nonegative localized potential centered at lattice site x_n, and the η_n are i.i.d. random variables uniformly distributed in $[0, 1]$.

Operators of the type given by (4.8) are known to have certain properties – see, e.g. [72]. If one first chooses a specific energy, then for a.e. realization V_ω of the random potential, that energy is part of its continuous spectrum (i.e., is not an eigenvalue of $\mathcal{H}_\omega$). On the other hand, if one first chooses a specific realization V_ω, then there is almost surely a (countable) dense set of eigenvalues (in some appropriate energy interval) of $\mathcal{H}_\omega$, and of course this set depends on ω.

Returning to the EA model in (say) three dimensions, it is amusing to speculate that a phenomenon like that described above for localization might occur for EA spin glasses, as follows (a) For every arbitrarily chosen T, there is a unique Gibbs state for a.e. $\mathcal{J}$ but (b) there exists a (deterministic) T_c, such that for a.e. $\mathcal{J}$, the set of temperatures T such that there are multiple Gibbs states (e.g., with $q_{EA} \neq 0$) for that $\mathcal{J}$ is dense in the entire temperature interval $[0, T_c]$. By an application of Fubini's theorem [73], property (a) would necessarily imply that the set of such $T(\mathcal{J})$'s would have zero Lebesgue measure in the temperature line.

Acknowledgments

This research was partially supported by NSF Grants DMS-01-02587 (CMN) and DMS-01-02541 (DLS). C.M. Newman was partially supported by the

174   Charles M. Newman and Daniel L. Stein

National Science Foundation under grant DMS-01-02587. D.L. Stein was partially supported by the National Science Foundation under grant DMS-01-02541.

References

1. D. Sherrington and S. Kirkpatrick, *Phys. Rev. Lett.* **35**, 1792 (1975)
2. S. Edwards and P.W. Anderson, *J. Phys. F* **5**, 965 (1975)
3. G. Parisi, *Phys. Rev. Lett.* **43**, 1754 (1979)
4. G. Parisi, *Phys. Rev. Lett.* **50**, 1946 (1983)
5. M. Mézard, G. Parisi, N. Sourlas, G. Toulouse, and M. Virasoro, *Phys. Rev. Lett.* **52**, 1156 (1984)
6. M. Mézard, G. Parisi, N. Sourlas, G. Toulouse, and M. Virasoro, *J. Phys. (Paris)* **45**, 843 (1984)
7. We gloss over here the serious problems surrounding the definition of pure states in the SK model. See the companion paper in this volume [21] for a more thorough discussion
8. C.M. Newman and D.L. Stein, *J. Phys.: Cond. Matter* **15**, R1319 (2003)
9. C.M. Newman and D.L. Stein, *J. Stat. Phys.* **106**, 213 (2002)
10. H.O. Georgii, *Gibbs Measures and Phase Transitions* (de Gruyter Studies in Mathematics, Berlin, 1988)
11. D. Ruelle, *Statistical Mechanics* (Benjamin, New York, 1969)
12. O. Lanford, in *Statistical Mechanics and Mathematical Problems, Lecture Notes in Physics, v. 20* (Springer, Berlin Heidelberg New York 1973), pp. 1–135
13. B. Simon, *The Statistical Mechanics of Lattice Gases v. 1*, (Princeton University Press, Princeton, 1993)
14. J. Slawny, in C. Domb and J.L. Lebowitz (eds.), *Phase Transitions and Critical Phenomena v. 11* (Academic, London, 1986), pp. 128–205
15. R.L. Dobrushin and S.B. Shlosman, in S.P. Novikov (ed.), *Soviet Scientific Reviews C. Math. Phys., v. 5* (Harwood Academic, New York, 1985), pp. 53–196
16. A.C.D. van Enter and J.L. van Hemmen, *Phys. Rev. A* **29**, 355 (1984)
17. K. Binder and A.P. Young, *Rev. Mod. Phys.* **58**, 801 (1986)
18. M. Mézard, G. Parisi, and M.A. Virasoro, *Spin Glass Theory and Beyond* (World Scientific, Singapore, 1987)
19. C.M. Newman and D.L. Stein, *Phys. Rev. B* **46**, 973 (1992)
20. C.M. Newman and D.L. Stein, *Phys. Rev. Lett.* **76**, 4821 (1996)
21. C.M. Newman and D.L. Stein, Local vs. Global Variables for Spin Glasses, p. 145, this volume
22. C.M. Newman and D.L. Stein, in *Mathematics of Spin Glasses and Neural Networks*, ed. A. Bovier and P. Picco (Birkhäuser, Boston, 1997), pp. 243–287
23. C.M. Newman and D.L. Stein, *Phys. Rev. E* **55**, 5194 (1997)
24. C.M. Newman and D.L. Stein, *Phys. Rev. E* **57**, 1356 (1998)
25. M. Aizenman and J. Wehr, *Commun. Math. Phys.* **130**, 489 (1990)
26. F. Krzakala and O.C. Martin, *Phys. Rev. Lett.* **85**, 3013 (2000)
27. M. Palassini and A.P. Young, *Phys. Rev. Lett.* **85**, 3017 (2000)
28. C.M. Newman and D.L. Stein, *Phys. Rev. Lett.* **87**, 077201 (2001)
29. C.M. Newman and D.L. Stein, *Annales Henri Poincaré* **4**, Suppl. 1, S497 (2003)
30. A.T. Ogielski, *Phys. Rev. B* **32**, 7384 (1985)

31. A.T. Ogielski and I. Morgenstern, *Phys. Rev. Lett.* **54**, 928 (1985)
32. N. Kawashima and A.P. Young, *Phys. Rev. B* **53**, R484 (1996)
33. M.E. Fisher and R.R.P. Singh, G. Grimmett and D.J.A. Welsh (eds.) in *Disorder in Physical Systems*, (Clarendon, Oxford, 1990), pp. 87–111
34. M.J. Thill and H.J. Hilhorst, *J. Phys. I* **6**, 67 (1996)
35. C.M. Newman and D.L. Stein, *Phys. Rev. Lett.* **84**, 3966 (2000)
36. C.M. Newman and D.L. Stein, *Phys. Rev. E* **63**, 16101-1 (2001)
37. W.L. McMillan, *J. Phys. C* **17**, 3179 (1984)
38. A.J. Bray and M.A. Moore, *Phys. Rev. B* **31**, 631 (1985)
39. A.J. Bray and M.A. Moore, *Phys. Rev. Lett.* **58**, 57 (1987)
40. D.S. Fisher and D.A. Huse, *Phys. Rev. Lett.* **56**, 1601 (1986)
41. D.A. Huse and D.S. Fisher, *J. Phys. A* **20**, L997 (1987)
42. D.S. Fisher and D.A. Huse, *J. Phys. A* **20**, L1005 (1987)
43. D.S. Fisher and D.A. Huse, *Phys. Rev. B* **38**, 386 (1988)
44. C.M. Newman and D.L. Stein, *Phys. Rev. Lett.* **76**, 515 (1996)
45. S. Franz, G. Parisi, and M.A. Virasoro, *J. Phys. I (France)* **4**, 1657 (1994)
46. E. Marinari, G. Parisi, F. Ricci-Tersenghi, J.J. Ruiz-Lorenzo, and F. Zuliani, *J. Stat. Phys.* **98**, 973 (2000)
47. C.M. Newman and D.L. Stein, *Phys. Rev. Lett.* **72**, 2286 (1994)
48. C.M. Newman and D.L. Stein, *J. Stat. Phys.* **82**, 1113 (1996)
49. J.R. Banavar, M. Cieplak, A. Maritan, *Phys. Rev. Lett.* **72**, 2320 (1994)
50. S. Franz, M. Mézard, G. Parisi, and L. Peliti, *Phys. Rev. Lett.* **81**, 1758 (1998)
51. P.W. Kasteleyn and C.M. Fortuin, *J. Phys. Soc. Jpn.* **26**, 11 (1969)
52. C.M. Fortuin and P.W. Kasteleyn, *Physica* **57**, 536 (1972)
53. A. Gandolfi, M.S. Keane, and C.M. Newman, *Prob. Theor. Rel. Fields* **92**, 511 (1992)
54. R.H. Swendsen and J.S. Wang, *Phys. Rev. Lett.* **58**, 86 (1987)
55. Y. Kasai and A. Okiji, *Prog. Theor. Phys.* **79**, 1080 (1988)
56. C. Newman, in *Probability and Phase Transitions*, edited by G. Grimmett (Kluwer, Dordrecht, 1993), pp. 247–260
57. G.R. Grimmett, *Percolation* (Springer, Berlin Heidelberg New York, 1999)
58. C.M. Fortuin, P.W. Kasteleyn, and J. Ginibre, *Commun. Math. Phys.* bf 22, 89 (1971)
59. T.E. Harris, *Proc. Camb. Philos. Soc.* **56**, 13 (1960)
60. M.F. Sykes and J.W. Essam, *Phys. Rev. Lett.* **10**, 3 (1963)
61. J. Wierman, *Adv. Appl. Prob.* **13**, 293 (1981)
62. J.M. Kosterlitz and D.J. Thouless, *J. Phys. C* **6**, 1181 (1973)
63. J.R. Banavar and A.J. Bray, *Phys. Rev. B* **35**, 8888 (1987)
64. I. Kondor, *J. Phys. A, Math. Gen.* **22**, L163 (1989)
65. M. Ney-Nifle and H.J. Hilhorst, *Physica A* **193**, 48 (1993)
66. F. Ritort, *Phys. Rev. B* **50**, 6844 (1994)
67. A. Billoire and E. Marinari, *J. Phys. A, Math. Gen.* **33**, L265 (2000)
68. J-P. Bouchaud, V. Dupuis, J. Hammann, and E. Vincent, *Phys. Rev. B* **65**, 024439 (2002)
69. A. Billoire and E. Marinari, *Europhys. Lett.* **60**, 775 (2002)
70. T. Rizzo and A. Crisanti, *Phys. Rev. Lett.* **90**, 137201 (2003)
71. M. Aizenman, A. Elgart, S. Naboko, J.H. Schenker, and G. Stolz, *Invent. Math.* **163**, 343 (2006)
72. R. Carmona and J. Lacroix, *Spectral Theory of Random Schrödinger Operators* (Birkhäuser, Boston, 1990)
73. W. Feller, *An Introduction to Probability Theory and Its Applications, Vol. II* (Wiley, New York, 1971), p. 124

Index

Lecture Notes in Mathematics

For information about earlier volumes
please contact your bookseller or Springer
LNM Online archive: springerlink.com

Vol. 1807: V. D. Milman, G. Schechtman (Eds.), Geometric Aspects of Functional Analysis. Israel Seminar 2000-2002 (2003)

Vol. 1808: W. Schindler, Measures with Symmetry Properties (2003)

Vol. 1809: O. Steinbach, Stability Estimates for Hybrid Coupled Domain Decomposition Methods (2003)

Vol. 1810: J. Wengenroth, Derived Functors in Functional Analysis (2003)

Vol. 1811: J. Stevens, Deformations of Singularities (2003)

Vol. 1812: L. Ambrosio, K. Deckelnick, G. Dziuk, M. Mimura, V. A. Solonnikov, H. M. Soner, Mathematical Aspects of Evolving Interfaces. Madeira, Funchal, Portugal 2000. Editors: P. Colli, J. F. Rodrigues (2003)

Vol. 1813: L. Ambrosio, L. A. Caffarelli, Y. Brenier, G. Buttazzo, C. Villani, Optimal Transportation and its Applications. Martina Franca, Italy 2001. Editors: L. A. Caffarelli, S. Salsa (2003)

Vol. 1814: P. Bank, F. Baudoin, H. Föllmer, L.C.G. Rogers, M. Soner, N. Touzi, Paris-Princeton Lectures on Mathematical Finance 2002 (2003)

Vol. 1815: A. M. Vershik (Ed.), Asymptotic Combinatorics with Applications to Mathematical Physics. St. Petersburg, Russia 2001 (2003)

Vol. 1816: S. Albeverio, W. Schachermayer, M. Talagrand, Lectures on Probability Theory and Statistics. Ecole d'Eté de Probabilités de Saint-Flour XXX-2000. Editor: P. Bernard (2003)

Vol. 1817: E. Koelink, W. Van Assche(Eds.), Orthogonal Polynomials and Special Functions. Leuven 2002 (2003)

Vol. 1818: M. Bildhauer, Convex Variational Problems with Linear, nearly Linear and/or Anisotropic Growth Conditions (2003)

Vol. 1819: D. Masser, Yu. V. Nesterenko, H. P. Schlickewei, W. M. Schmidt, M. Waldschmidt, Diophantine Approximation. Cetraro, Italy 2000. Editors: F. Amoroso, U. Zannier (2003)

Vol. 1820: F. Hiai, H. Kosaki, Means of Hilbert Space Operators (2003)

Vol. 1821: S. Teufel, Adiabatic Perturbation Theory in Quantum Dynamics (2003)

Vol. 1822: S.-N. Chow, R. Conti, R. Johnson, J. Mallet-Paret, R. Nussbaum, Dynamical Systems. Cetraro, Italy 2000. Editors: J. W. Macki, P. Zecca (2003)

Vol. 1823: A. M. Anile, W. Allegretto, C. Ringhofer, Mathematical Problems in Semiconductor Physics. Cetraro, Italy 1998. Editor: A. M. Anile (2003)

Vol. 1824: J. A. Navarro González, J. B. Sancho de Salas, $\mathscr{C}^\infty$ – Differentiable Spaces (2003)

Vol. 1825: J. H. Bramble, A. Cohen, W. Dahmen, Multiscale Problems and Methods in Numerical Simulations, Martina Franca, Italy 2001. Editor: C. Canuto (2003)

Vol. 1826: K. Dohmen, Improved Bonferroni Inequalities via Abstract Tubes. Inequalities and Identities of Inclusion-Exclusion Type. VIII, 113 p, 2003.

Vol. 1827: K. M. Pilgrim, Combinations of Complex Dynamical Systems. IX, 118 p, 2003.

Vol. 1828: D. J. Green, Gröbner Bases and the Computation of Group Cohomology. XII, 138 p, 2003.

Vol. 1829: E. Altman, B. Gaujal, A. Hordijk, Discrete-Event Control of Stochastic Networks: Multimodularity and Regularity. XIV, 313 p, 2003.

Vol. 1830: M. I. Gil', Operator Functions and Localization of Spectra. XIV, 256 p, 2003.

Vol. 1831: A. Connes, J. Cuntz, E. Guentner, N. Higson, J. E. Kaminker, Noncommutative Geometry, Martina Franca, Italy 2002. Editors: S. Doplicher, L. Longo (2004)

Vol. 1832: J. Azéma, M. Émery, M. Ledoux, M. Yor (Eds.), Séminaire de Probabilités XXXVII (2003)

Vol. 1833: D.-Q. Jiang, M. Qian, M.-P. Qian, Mathematical Theory of Nonequilibrium Steady States. On the Frontier of Probability and Dynamical Systems. IX, 280 p, 2004.

Vol. 1834: Yo. Yomdin, G. Comte, Tame Geometry with Application in Smooth Analysis. VIII, 186 p, 2004.

Vol. 1835: O.T. Izhboldin, B. Kahn, N.A. Karpenko, A. Vishik, Geometric Methods in the Algebraic Theory of Quadratic Forms. Summer School, Lens, 2000. Editor: J.-P. Tignol (2004)

Vol. 1836: C. Năstăsescu, F. Van Oystaeyen, Methods of Graded Rings. XIII, 304 p, 2004.

Vol. 1837: S. Tavaré, O. Zeitouni, Lectures on Probability Theory and Statistics. Ecole d'Eté de Probabilités de Saint-Flour XXXI-2001. Editor: J. Picard (2004)

Vol. 1838: A.J. Ganesh, N.W. O'Connell, D.J. Wischik, Big Queues. XII, 254 p, 2004.

Vol. 1839: R. Gohm, Noncommutative Stationary Processes. VIII, 170 p, 2004.

Vol. 1840: B. Tsirelson, W. Werner, Lectures on Probability Theory and Statistics. Ecole d'Eté de Probabilités de Saint-Flour XXXII-2002. Editor: J. Picard (2004)

Vol. 1841: W. Reichel, Uniqueness Theorems for Variational Problems by the Method of Transformation Groups (2004)

Vol. 1842: T. Johnsen, A.L. Knutsen, K3 Projective Models in Scrolls (2004)

Vol. 1843: B. Jefferies, Spectral Properties of Noncommuting Operators (2004)

Vol. 1844: K.F. Siburg, The Principle of Least Action in Geometry and Dynamics (2004)

Vol. 1845: Min Ho Lee, Mixed Automorphic Forms, Torus Bundles, and Jacobi Forms (2004)

Vol. 1846: H. Ammari, H. Kang, Reconstruction of Small Inhomogeneities from Boundary Measurements (2004)

Vol. 1847: T.R. Bielecki, T. Björk, M. Jeanblanc, M. Rutkowski, J.A. Scheinkman, W. Xiong, Paris-Princeton Lectures on Mathematical Finance 2003 (2004)

Vol. 1848: M. Abate, J. E. Fornaess, X. Huang, J. P. Rosay, A. Tumanov, Real Methods in Complex and CR Geometry, Martina Franca, Italy 2002. Editors: D. Zaitsev, G. Zampieri (2004)

Vol. 1849: Martin L. Brown, Heegner Modules and Elliptic Curves (2004)

Vol. 1850: V. D. Milman, G. Schechtman (Eds.), Geometric Aspects of Functional Analysis. Israel Seminar 2002-2003 (2004)

Vol. 1851: O. Catoni, Statistical Learning Theory and Stochastic Optimization (2004)

Vol. 1852: A.S. Kechris, B.D. Miller, Topics in Orbit Equivalence (2004)

Vol. 1853: Ch. Favre, M. Jonsson, The Valuative Tree (2004)

Vol. 1854: O. Saeki, Topology of Singular Fibers of Differential Maps (2004)

Vol. 1855: G. Da Prato, P.C. Kunstmann, I. Lasiecka, A. Lunardi, R. Schnaubelt, L. Weis, Functional Analytic Methods for Evolution Equations. Editors: M. Iannelli, R. Nagel, S. Piazzera (2004)

Vol. 1856: K. Back, T.R. Bielecki, C. Hipp, S. Peng, W. Schachermayer, Stochastic Methods in Finance, Bres-

Recent Reprints and New Editions

4. Manuscripts should in general be submitted in English. Final manuscripts should contain at least 100 pages of mathematical text and should always include

 – a general table of contents;
 – an informative introduction, with adequate motivation and perhaps some historical remarks: it should be accessible to a reader not intimately familiar with the topic treated;
 – a global subject index: as a rule this is genuinely helpful for the reader.

Lecture Notes volumes are, as a rule, printed digitally from the authors' files. We strongly recommend that all contributions in a volume be written in the same LaTeX version, preferably LaTeX2e. To ensure best results, authors are asked to use the LaTeX2e style files available from Springer's web-server at

ftp://ftp.springer.de/pub/tex/latex/mathegl/mono.zip (for monographs) and
ftp://ftp.springer.de/pub/tex/latex/mathegl/mult.zip (for summer schools/tutorials).

Additional technical instructions, if necessary, are available on request from:

lnm@springer-sbm.com.

5. Careful preparation of the manuscripts will help keep production time short besides ensuring satisfactory appearance of the finished book in print and online. After acceptance of the manuscript authors will be asked to prepare the final LaTeX source files (and also the corresponding dvi-, pdf- or zipped ps-file) together with the final printout made from these files. The LaTeX source files are essential for producing the full-text online version of the book. For the existing online volumes of LNM see:
http://www.springerlink.com/openurl.asp?genre=journal&issn=0075-8434.

The actual production of a Lecture Notes volume takes approximately 8 weeks.

6. Volume editors receive a total of 50 free copies of their volume to be shared with the authors, but no royalties. They and the authors are entitled to a discount of 33.3 % on the price of Springer books purchased for their personal use, if ordering directly from Springer.

7. Commitment to publish is made by letter of intent rather than by signing a formal contract. Springer-Verlag secures the copyright for each volume. Authors are free to reuse material contained in their LNM volumes in later publications: A brief written (or e-mail) request for formal permission is sufficient.

Addresses:

Professor J.-M. Morel, CMLA,
École Normale Supérieure de Cachan,
61 Avenue du Président Wilson, 94235 Cachan Cedex, France
E-mail: Jean-Michel.Morel@cmla.ens-cachan.fr

Professor F. Takens, Mathematisch Instituut,
Rijksuniversiteit Groningen, Postbus 800,
9700 AV Groningen, The Netherlands
E-mail: F.Takens@math.rug.nl

Professor B. Teissier, Institut Mathématique de Jussieu,
UMR 7586 du CNRS, Équipe "Géométrie et Dynamique",
175 rue du Chevaleret, 75013 Paris, France
E-mail: teissier@math.jussieu.fr

For the "Mathematical Biosciences Subseries" of LNM :
Professor P. K. Maini, Center for Mathematical Biology,
Mathematical Institute, 24-29 St Giles,
Oxford OX1 3LP, UK
E-mail : maini@maths.ox.ac.uk

Springer, Mathematics Editorial I, Tiergartenstr. 17,
69121 Heidelberg, Germany,
Tel.: +49 (6221) 487-8410
Fax: +49 (6221) 487-8355
E-mail: lnm@springer-sbm.com